World Scientific Series on Public Policy and Technological Innovation – Vol. 1

Student Start-Ups

The New Landscape of Academic Entrepreneurship

World Scientific Series on Public Policy and Technological Innovation

Series Editor-in-Chief: Donald Siegel (*Arizona State University*)

Published

Vol. 1   *Student Start-Ups: The New Landscape of Academic Entrepreneurship*
by Mike Wright, Philippe Mustar & Donald Siegel

World Scientific Series on Public Policy and Technological Innovation – Vol. 1

Student Start-Ups

The New Landscape of Academic Entrepreneurship

Mike Wright
Imperial College Business School, UK

Philippe Mustar
MINES ParisTech-PSL, France

Donald Siegel
Arizona State University, USA

World Scientific

NEW JERSEY · LONDON · SINGAPORE · BEIJING · SHANGHAI · HONG KONG · TAIPEI · CHENNAI · TOKYO

Published by

World Scientific Publishing Co. Pte. Ltd.

5 Toh Tuck Link, Singapore 596224

USA office: 27 Warren Street, Suite 401-402, Hackensack, NJ 07601

UK office: 57 Shelton Street, Covent Garden, London WC2H 9HE

Library of Congress Cataloging-in-Publication Data
Names: Wright, Mike, 1952–   author. | Mustar, Philippe, author. |
    Siegel, Donald S., 1959–   author.
Title: Student start-ups : the new landscape of academic entrepreneurship /
    Mike Wright, Philippe Mustar, Donald Siegel.
Description: Singapore ; Hackensack, NJ : World Scientific, 2020. | Series: World scientific series on
    public policy and technological innovation; vol. 1 | Includes bibliographical references and index.
Identifiers: LCCN 2019035966 | ISBN 9789811208102 (hardcover) | ISBN 9789811214974 (pbk)
Subjects: LCSH: Entrepreneurship--Study and teaching (Higher) | New business enterprises--
    Study and teaching (Higher) | Education, Higher--Economic aspects.
Classification: LCC HB615 .W69 2020 | DDC 658.1/10711--dc23
LC record available at https://lccn.loc.gov/2019035966

British Library Cataloguing-in-Publication Data
A catalogue record for this book is available from the British Library.

For any available supplementary material, please visit
https://www.worldscientific.com/worldscibooks/10.1142/11494#t=suppl

Desk Editors: Dr. Sree Meenakshi Sajani/Sandhya Venkatesh

Typeset by Stallion Press
Email: enquiries@stallionpress.com

Printed in Singapore

Preface

There has been a substantial rise in the number of entrepreneurship courses and programs at colleges and universities. The growth in undergraduate entrepreneurship has been especially rapid. Despite the rise of student entrepreneurship, there have been only a few academic studies of this phenomenon. Thus, we know little about the antecedents and consequences of these activities. We hope to fill this gap in our book, by presenting the best available quantitative data, along with detailed case studies of student entrepreneurship programs. Our objective is to synthesize this evidence to provide a structured approach to understanding the development of this phenomenon, as well as to generate insights and implications for practice, policy and research.

We are indebted to many individuals who provided quantitative and qualitative data about entrepreneurial programs and initiatives, most notably Andy Goldstein, Jonas van Hove, Harveen Chugh, Chris Corbishley, Simon Mosey, Vangelis Souitaris, Riccardo Fini and Rosa Grimaldi. We are also grateful to the professors, administrators and students at the Cambridge University, Cass Business School, King's College London, University College London, Aalto University, MINES ParisTech-PSL, the Foundation Mines ParisTech, the University at Albany, SUNY and Arizona State University. Finally, we thank Martin Kenney for his encouragement and extensive suggestions and comments on our earlier drafts.

About the Authors

Mike Wright is Professor of Entrepreneurship and founding Director of the Center for Management Buyout Research at Imperial College Business School. He is the editor of *Foundations and Trends in Entrepreneurship* and a former co-editor of the *Journal of Management Studies, Strategic Entrepreneurship Journal, Academy of Management Perspectives* and the *Journal of Technology Transfer*. He is Chair of the Society for the Advancement of Management Studies and is a Fellow of the British Academy and of the Strategic Management Society. He has been ranked #1 in the world for research on academic entrepreneurship.

Philippe Mustar is Professor of Entrepreneurship and Head of the Entrepreneurship Unit (POLLEN) at MINES ParisTech — PSL Research University. He is a visiting Professor at the Imperial College Business School and member of the Editorial Boards of *Research Policy* and *Technovation*. He is a pioneer of research in academic entrepreneurship. He advises start-ups and international organizations in innovation and entrepreneurship policy. His ongoing research interests are in student start-ups and entrepreneurship education.

Donald Siegel is Foundation Professor of Public Policy and Management and Director of the School of Public Affairs at Arizona State University. He is an editor of the *Journal of Management Studies* and the *Journal of Technology Transfer*, a former co-editor of *Academy of Management Perspectives* and a former Dean of the School of Business at the University at Albany, SUNY, where he implemented numerous programs to support student entrepreneurship and also raised substantial funds dedicated to this activity. In 2016, he was elected as a Fellow of the Academy of Management. He has been ranked #2 in the world for research on academic entrepreneurship.

Contents

List of Figures

List of Tables

Chapter 4

Chapter 6

Chapter 1

The Growing Phenomenon of Student Entrepreneurship

Introduction

Three of the five most valuable companies in the world were started by students. Larry Page and Steve Brin, co-founders of Google, left Stanford's Ph.D. program to launch the project that subsequently became Google. Bill Gates dropped out of Harvard College without completing his undergraduate degree in math or computer science, in order to start Microsoft. Mark Zuckerberg launched Facebook from his dormitory room at Harvard. Apple was started by an itinerant student, Steve Jobs, who eventually dropped out of Reed College. Only Amazon does not have strong ties to students. Among other major corporations, Michael Dell, founder of the eponymous computer corporation and Yahoo! founder Jerry Yang started ventures when they were students.

These examples are just the pinnacle of an entrepreneurial movement that has seen a substantial rise in university-based start-ups, both by faculty and especially, by students. This student start-up movement is associated with rising global demand for entrepreneurship educational programs and entrepreneurship support structures at universities, which has called into question traditional teaching approaches and radically altered such teaching methods (OECD, 2015a).

Thus, there is growing interest in designing and evaluating student entrepreneurship programs and initiatives, from both managerial and public policy standpoints. Administrators are interested in spawning new firms because they can yield three benefits to the university: (1) revenue, (2) start-ups that make the university more attractive to current or prospective faculty members and students and (3) enhancing local/regional economic development. From a policy standpoint, in many countries, entrepreneurship has become an attractive employment option for many university graduates faced with a decline in traditional career options in corporations. This is especially true in Europe, where job growth has been sluggish since the global financial crisis of 2008.

Given these trends, we were inspired to write this book after many years of studying university entrepreneurship. Most research on university entrepreneurship has focused on start-ups launched by faculty. When students have been involved in such efforts, the focus has been on the role of *doctoral* students and *postdocs* in faculty-led ventures (Stephan, 2009). It turns out that most university-based start-ups are not launched by faculty. The most detailed study of university-based start-ups at leading research universities was conducted by Boh *et al.* (2016). The authors presented detailed case studies of university-based start-ups at the following institutions: Harvard, MIT, Stanford, University of Arizona, UC Berkeley, University of Maryland, University of North Carolina and the University of Utah. They report that 54 percent of these firms involved Ph.D. students or postdocs, with 13 percent also involving business school students. Another interesting stylized fact is that 23 percent of university-based start-ups were launched by students only, involving a master's or doctoral student and a business school student. Note that these were technology-based start-ups only, and thus, probably understate the true extent of student entrepreneurship, since each of these start-ups began with an invention disclosure to a university technology transfer office.

In addition to current students, many start-ups are launched by young alumni, i.e., recent university graduates. Even if these individuals were not entrepreneurs while at the university, the entrepreneurship programs they follow, the societies they join and the interactions they have with entrepreneurial culture prepare them for entrepreneurship later. Thus, we include recent graduates, given that it may take some time for entrepreneurial ideas that begin to be formulated at universities to become actual start-ups.

There is some debate regarding the appropriate cut-off point for regarding alumni as recent graduates. However, approximately 3 years after graduation appears to be the optimal choice. Nabi *et al.* (2017) demonstrate that the most important impact of entrepreneurship education leading to student start-ups is likely to occur post-program. Further, the evolving policy seems to recognize this issue (Council for Science and Technology, 2016) and the limited data availability reflects this basis (HESA, 2017).

Given that many research universities have adopted explicit strategies of promoting "technology transfer" (Siegel and Wessner, 2012), i.e., the commercialization of university-based research, as a form of regional economic development, it makes sense to include all of these ventures in the definition of student entrepreneurship. While most academic studies of university entrepreneurship have emphasized the commercialization of technology through start-ups based on formal intellectual property (IP) emanating from the university technology transfer office, this is a narrow view, especially in the context of student entrepreneurship. Most students, especially undergraduate students, are not working with professors or postdocs on federally funded research or even other types of sponsored research. Across the spectrum from elite research universities to non-elite colleges, students from all disciplines, including the social sciences, business administration, arts and humanities, are engaging in entrepreneurial activity that does not necessarily depend on IP, formal or informal, generated by the university. Accordingly, we encompass the broad range of high-tech and non-high-tech sectors.

Given that student entrepreneurship on college campuses is relatively embryonic, there are limited data on these start-ups. That is, both universities and the government have traditionally not systematically tracked this activity. Also, theoretical frameworks for studying this activity are also embryonic. In this book, we will attempt to fill this void by providing such frameworks. We also present voluminous qualitative evidence on key programs designed to stimulate student entrepreneurship across the globe. In addition, we consider how barriers to the development of student entrepreneurship can be overcome.

The best available data relate to start-ups that have received venture funding. For many years, venture capitalists have targeted universities for investment. For example, the first major venture capital firm, American Research and Development (ARD), was founded in 1946 by Georges Doriot, the former Dean of the Harvard Business School, along with Karl Compton, the former President of MIT. ARD was launched with a focus on MIT technology-based start-ups.

The limited available data strongly suggest that the extent of this activity is significant and international. For example, the most active U.S. research universities in promoting mostly tech-based student entrepreneurship are Stanford, UC Berkeley, MIT and Harvard, which reported a total of 3,578 recent alumni entrepreneurs receiving venture-backed funding, according to latest data for 2016/2017 (Table 1). In the U.K., from 2001 to 2016, the annual number of start-ups generated by recent alumni increased dramatically to around 4,000 per year (Table 3). In both the U.K. and France, the number of start-ups per year by students and young graduates is approximately 20 times greater than the number created by university faculty (HESA, 2017). This activity is not confined to developed economies; there is evidence that continents with emerging economies are also promoting student entrepreneurship.

To shed more light on what is happening behind these headline figures, we begin by presenting quantitative and qualitative evidence on the extent and nature of student start-ups at universities in the U.S., Europe and Africa.

The Extent and Nature of Student Entrepreneurship

In recent years, universities have started improving their data collection on student entrepreneurship, as the start-up phenomenon has become more prevalent on campus. As these efforts mature, we will eventually have better data. In the meantime, we have attempted to gather the best quantitative and qualitative data on student start-ups from several nations. This data is not meant to be a comprehensive global survey of student entrepreneurship but rather an indicator of the extent and variety in the development of this activity. As you will see, these countries have vastly different university governance structures and varying approaches to promoting student entrepreneurship.

Table 1:   Top 20 undergraduate institutions for generating venture-backed start-ups

Rank	University	Entrepreneurs receiving venture-backed funding	Companies receiving venture-backed funding	Venture capital raised ($ million)
1	Stanford	1006	850	$18,146
2	UC-Berkeley	997	881	$14,239
3	MIT	813	695	$12,874
4	Harvard	762	673	$17,204
5	University of Pennsylvania	724	648	$9,475
6	Cornell	635	585	$10,777
7	University of Michigan	607	546	$7,767
8	University of Texas	561	511	$4,763
9	Tel Aviv	515	429	$5,101
10	University of Illinois	451	415	$5,462
11	UCLA	432	406	$6,988
12	Yale	421	379	$7,449
13	Princeton	408	382	$6,976
14	University of Wisconsin	397	350	$2,632
15	USC	381	341	$3,476
16	Technion — Israel Institute of Technology	379	323	$4,765
17	Carnegie Mellon	378	324	$4,592
18 (tie)	Columbia	373	347	$4,995
19 (tie)	Brown	373	338	$6,426
20	University of Waterloo	361	275	$5,067

Source: Pitchbook Universities Report, 2016–2017 edition.

For instance, the U.S. has many different types of higher educational institutions. There are major state university systems, such as the State University of New York, which has 64 campuses, including community colleges (e.g., Hudson Valley Community College), comprehensive colleges (e.g., SUNY Geneseo), technology colleges (e.g., Fashion Institute of Technology), university research centers (e.g., University at Albany) and other doctoral-granting institutions (e.g., SUNY Downstate Medical Center). The U.S. also has many private colleges and universities, including the top research universities in the world (e.g., Harvard, MIT, Yale, Princeton, Stanford and Columbia), liberal arts colleges (e.g., Amherst College and Williams College) as well as institutions with a religious focus or mission (e.g., Jesuit colleges, such as Boston College; Baptist colleges, such as Baylor University; and Christian Evangelical colleges, such as Oral Roberts University). Outside the U.S., most university systems are dominated by public institutions. In these nations, private colleges and universities are rare. These countries also vary in terms of the extent of centralization of approaches to student entrepreneurship.

United States

In the U.S., some of the richest data on student entrepreneurship are provided by PitchBook. Note that this is not a "pure" measure of student entrepreneurship, since it includes information on start-ups launched by recent alumni, as previously defined in this chapter. PitchBook collects information on mergers and acquisitions, private equity and venture capital transactions, public and private companies, investors, funds, firms and entrepreneurs (especially, founders). Each year, PitchBook ranks the top universities, in terms of their production of venture capital-backed entrepreneurs, with detailed breakdowns of top programs by sector, exit rates and more in the "PitchBook Universities Report". They also track founders of companies who received first round of venture funding during the previous year.

Table 1 presents the latest results for the top 20 undergraduate institutions for generating venture-backed start-ups. Not surprisingly, the top four universities are located in the Bay Area (Stanford and UC Berkeley) and Boston/Cambridge (MIT and Harvard). However, proximity to venture capitalists is not the only factor. Schools that do well in this ranking include Cornell, Michigan, the University of Illinois Urbana–Champaign and the University of Waterloo. Table 2 presents data on the top 20 MBA programs, in terms of generating start-ups. Note that two Israeli

Table 2: Top 20 MBA programs in generating start-ups

Rank	University	Entrepreneurs receiving venture-backed funding	Companies receiving venture-backed funding	Venture capital raised ($ million)
1	Harvard	1069	961	$22,425
2	Stanford	720	636	$14,475
3	University of Pennsylvania	577	506	$10,602
4	Northwestern	409	381	$4,718
5	INSEAD	393	348	$6,131
6	MIT	379	336	$6,196
7	Columbia	377	352	$4,408
8	University of Chicago	363	330	$4,070
9	UC-Berkeley	300	272	$3,963
10	UCLA	227	212	$2,919
11	NYU	218	213	$2,954
12	Tel Aviv	187	178	$2,366
13	London Business School	182	164	$1,313
14	University of Texas	146	132	$1,253
15	University of Michigan	137	125	$1,093
16	Duke	135	131	$1,021
17	Babson	125	109	$1,440
18	Cornell	105	103	$1,137
19	USC	99	95	$1,107
20	Dartmouth	87	81	$1,019

Source: Pitchbook Universities Report, 2016–2017 edition.

universities appear on this list: Technion-Israel Institute of Technology and Tel Aviv University, which is consistent with the view that Israel is the "Start-up Nation" (Senor and Singer, 2009).

Several universities have started tracking the economic impact of student entrepreneurship, as part of an effort to document the overall economic impact of the university on regional economic development. In 2012, Charles Eesley and William F. Miller from Stanford University published "Stanford University's Economic Impact via Innovation and Entrepreneurship" (Eesley and Miller, 2012). This important study is based on a large-scale, systematic survey of Stanford alumni and faculty that goes beyond current students and recent graduates. The authors estimate that Stanford alumni and faculty have created 39,900 companies since the 1930s. These businesses include some of the world's most recognized companies like Google, Nike, Cisco, Hewlett-Packard, Charles Schwab, Yahoo!, Gap, VMware, IDEO, Netflix and Tesla. Several of these businesses were launched by Stanford students (especially, graduate students). The study also calculates that these companies have created an estimated 5.4 million jobs and generate annual world revenues of $2.7 trillion. These companies if gathered collectively into an independent nation would constitute the world's 10 largest economies.

The report also shows the rise of social innovation at the university. It estimates that in addition to 39,900 for-profit organizations, graduates and faculty have created more than 30,000 non-profit organizations. Stanford graduates have founded, built or led thousands of ventures in the area of social innovation. Among these, the best-known are Kiva, a microfinance organization, and The Special Olympics and Acumen Fund, which supports entrepreneurs in developing economies.

The survey shows some other interesting results as follows:

- Around 29 percent of respondents reported being entrepreneurs who founded an organization (for-profit or non-profit).
- About 32 percent of alumni described themselves as an investor, early employee or a board member in a start-up at some point in their careers.
- A total of 25 percent of faculty respondents (some of whom are also alumni) reported founding or incorporating a firm at some point in their careers.

Among survey respondents who became entrepreneurs in the past decade, 55 percent reported choosing to study at Stanford specifically because of its entrepreneurial environment.

Although the data in the report include start-ups by older alumni and faculty, it describes how the university creates an ecosystem that encourages networking, creativity and entrepreneurship across schools and disciplines. The university encourages students to become involved in researching and prototyping their ideas. For many years, Stanford has also provided education designed to encourage and develop entrepreneurs. The report states that, "The university began offering classes in small business and entrepreneurship as enrolment mushroomed after the Second World War. Today, it offers dozens of courses and programs that educate and support potential entrepreneurs".

Case study of a successful student venture in the U.S.: The Google search engine

Our first case study concerns a successful student venture at Stanford. According to Hart (2004), the story of Google began when two Stanford University graduate students in computer science, Sergey Brin and Lawrence Page, were working on the Stanford Digital Library Project (SDLP), which was funded by NSF (mainly), DARPA, NASA and several corporate partners. SDLP's chief goal was to develop a single, integrated and universal digital library, which would replace traditional collections of books. Brin also received an NSF graduate fellowship.

The two students developed a new algorithm, called PageRank, which formed the basis for a search engine, called BackRub (see Brin and Page (1998)). Brin and Page continued to work on their search engine at SDLP, using NSF-funded equipment. Soon after, they sought private investment to commercialize their new search engine, which became the company ultimately known as Google. According to Singer (2014), Google's search engine received a boost when an article appeared in *Salon.com* arguing that Google's search results were superior to its rivals. Soon, it become obvious to most consumers and firms that Google's search engine, which had been supported by federal R&D investment through NSF, DARPA and NASA, was superior to popular search engines, such as AltaVista, Yahoo!, Excite.com, Lycos, Netscape's Netcenter, AOL.com, Go.com and MSN.com.

Since establishing their dominance in the search engine market, the company has branched out into advertising, social networking, email hosting and operating systems for the mobile device market while continuing to improve and expand its search engine algorithms (e.g., Google Scholar, Google Maps), including several attempts to delve into social networking (e.g., Google+, Google Buzz, Google Friend Connect). Google's search engine has also created a whole new marketing industry centered on search engine optimization. It also serves as an enabling technology for numerous service industries. Today, Alphabet, Google's parent company, is a Fortune 100 company, generating $110.8 billion in revenue in 2017, and $12.7 billion in net profit.

United Kingdom

At U.K. universities, according to the Global Entrepreneurship Monitor (GEM) data (Hart and Bonner, 2015), recent alumni are somewhat more likely than current students and non-graduates to establish new firms. The GEM data reveal that the Total Early-Stage Entrepreneurship Activity (TEA) rate in 2014 (which includes both nascent and start-up entrepreneurs) was 10.8 percent for alumni, compared to the U.K. average of 8.6 percent and a rate of 7.4 percent for those who were not recent graduates. The GEM data also reveal that alumni are significantly more likely to have growth ambitions for their ventures, defined as expecting to employ more than 10 people and to grow at more than 50 percent within 5 years.

Data for the U.K. are presented in Table 3. As noted earlier, the annual number of start-ups generated by recent alumni increased almost 14-fold between 2001 and 2016, with the annual numbers hovering around 4,000 in the past 3 years. In contrast, there were 215 faculty spin-offs and start-ups according to HESA data. By 2016, nearly 10,000 alumni start-ups were still active. These ventures are also having a significant economic impact, employing 18,560 workers and generating approximately $500,000 in revenue in 2013/2014. Many of these ventures appear to have attractive growth prospects, drawing significant amounts of external investment. There is substantial annual fluctuation in the amount of funds received and there does not appear to be an upward trend in funding commensurate with the rise in the number of ventures created. This pattern may be indicative of the shortcomings in the ecosystem regarding the provision of finance for student start-ups, a challenge we return to in

Table 3: Start-ups generated by recent alumni in the U.K.

	Start-ups by recent alumni[b]	Total number of active firms started by alumni	Survived 3 years	Estimated employment	Estimated revenue (£000)	Total investment received (£000)
2001/2002	337	n.a.	278	2,263	66,890	n.a.
2002/2003	489	732	323	1,493	10,5200	n.a.
2003/2004	572	905	206	1,999	12,8567	n.a.
2004/2005	974	2,058	635	4,438	11,1861	n.a.
2005/2006	1,172	2,811	867	5,477	84,808	n.a.
2006/2007	n.a.	n.a.	n.a.	n.a.	n.a.	n.a.
2007/2008	1,961	3,970	1,323	6,297	145,418	33,983
2008/2009	2,045	4,053	1,667	8,014	137,060	50,474
2009/2010	2,357	5,064	1,948	9,704	225,175	8,437
2010/2011	2,848	6,413	2,602	11,914	272,655	88,782
2011/2012	2,726	7,151	2,824	13,617	345,999	31,627
2012/2013	3,502	8,163	3,270	15,588	376,407	28,544
2013/2014	4,603	9,963	3,873	18,560	474,667	74,305
2014/2015	4,160	10,956	4,474	20,880	648,985	302,740
2015/2016[a]	3,890	1,1361	5,423	22,592	626,790	131,695

Notes: [a]In 2015/2016, an additional 106 social enterprises were created by alumni; [b]Graduate start-ups are defined as companies formed within 2 years of graduation, where the graduate has received assistance from the HEI. They may or may not be IP-based and include various types of business, including commercial and social enterprises.
Source: Various HE BCI Reports.

Chapter 4, but it may also reflect a lumpiness in the volume of student start-ups requiring substantial finance to facilitate growth as many start-ups are in ICT sectors with lower needs.

Indeed, the average size of alumni start-ups is considerably lower than that for start-ups and spin-offs by faculty. For example, in 2013/2014 the average alumni start-up had 1.86 employees while the average faculty spin-off/start-up had 10 employees (HESA, 2015), which suggests that faculty start-ups have a larger economic impact. We lack evidence on whether this difference is due to faculty identifying more attractive growth opportunities or the fact that they are receiving more support. Nevertheless, significant number of start-ups by alumni survive for at least 3 years. The number of alumni start-ups surviving 3 or more years rose by 18.4 percent between 2012/2013 and 2013/2014.

The creation of student start-ups in the U.K. has been aided by several government-supported initiatives. One such program is the Graduate Entrepreneurship Project (GEP), overseen by 10 universities in Yorkshire. Launched in 2007, this project aims to provide enterprise and business start-up support for students and graduates. Funding has been provided from the European Regional Development Fund for two consecutive rounds (2007–2010 and 2011–2014). Support also includes mentoring and advice from a full-time advisor based at the university, and funds to procure specific expertise and regular visits from external subject specific advisors, proof of concept funds of up to £1,000 and start-up grants of up to £2,500, channeled via the partner institutions, an entrepreneurs Boot Camp comprising a 4-day residential program, providing participants with an opportunity to develop their business plans, supported by expert business advice on a range of issues, and university showcases, networking events and workshops focusing on commercial and entrepreneurial skills. Since 2007, the GEP has helped establish 278 businesses, which survived more than a year.

Case study of a successful student venture in the U.K.: Waterfox

We now provide an illustration of how such support can be essential in launching a student venture and positioning it to attract external financial capital. In 2011, GEP helped a 19-year-old University of York student Alex Kontos start a venture to commercialize Waterfox, a high-performance web browser he had been developing for 3 years. Through GEP, the University of York provided him with a range of support services, including business advice, a proof of concept grant of £1,000 and the opportunity to attend a Boot Camp. In 2014, Kontos won the prestigious Duke of York Young Entrepreneur of the Year Award (University of York, 2014). The browser had in excess of 6 million downloads by 2016 and at the time of writing is on release 52. As Waterfox hasn't generated revenues yet, Kontos has also developed a search engine, Storm, with investment from a venture capitalist. The aim is to tempt users away from Google and to make a social contribution through a small percentage of any purchase made with participating e-tailers being given to charitable organizations. However, it also needs the student to be fully committed.

France

In France, the Law on Innovation and Research (henceforth, LIR) was enacted in July 1999. LIR was designed to promote high-tech start-ups from universities and public research organizations (e.g., national laboratories). A key aspect of this legislation is that it established an amended status for faculty and other university researchers to enhance their ability to commercialize their research, through the creation of new firms.

It is important to note that in France, faculty and other university researchers, along with researchers at public research organizations, are civil servants (or have a similar status). This status was considered an obstacle precluding academics from creating or participating in the creation of a private company. The law therefore amended the civil servant status of researchers and academic researchers, thus enabling them to participate in the creation of a private firm based on their research. An academic can now take temporary leave from his/her functions, create a company and become an associate or a manager, or stay in his/her university and offer scientific assistance to the spin-off. He/she can acquire shares in this company. For each case, an approval given by a National Public Sector Ethics Committee is necessary (even though almost all the recommendations of the Commission are favorable).

Between 2000 and 2014, approximately 180 academics sought and received authorization from the National Ethics Commission to participate in a new venture as executives (i.e., 12 per year), and 1,200 academics were authorized to serve as scientific consultants (i.e., 80 per year). Most of these academics were involved with start-ups that they helped create.[1] Allowing for the fact that several academics often participated in

---

[1] These data were collected every year in the annual activity reports of the National Public Sector Ethics Committee.

the same project, close to 800 firms were concerned. In sum, during these 15 years, about 55 firms were created annually as start-ups from research by a public laboratory, *and* involved full-time (CEO) or part-time (scientific consultant) research staff who came under the law on research and innovation (that is, tenured staff of universities, engineering schools and most public research organizations). This figure does not include academic start-ups (ASOs) created without one of the founders applying for authorization from their institution and from the ethics commission. However, these ASOs amount to no more than a maximum of 100 per year, even taking a very broad definition of an ASO.[2]

In France, public policy efforts to promote student entrepreneurs began in the early 2000s (see Chapter 6 for discussion of the development of these policies). In the mid-2000s, the French government developed early support measures for students with entrepreneurial projects. Unfortunately, there is no systematic data on the number of companies created each year by French students or recent alumni, in the aftermath of these programs. However, a recent policy initiative implemented by the French Ministry for Higher Education and Research allows us to estimate the number of student start-ups in France. This initiative confers the status of "student entrepreneur" on students and young alumni who have an entrepreneurial project.

This national status of student entrepreneur allows students and young graduates to launch start-ups with the support of a specific type of business incubator known in French as a *Pôle Etudiants pour l'Innovation, le Transfert et l'Entrepreneuriat* (PEPITE), that is, a Student Center for Innovation, Transfer and Entrepreneurship. There are currently 30 PEPITEs in France. They bring together universities, schools, government agencies, local entrepreneurs, venture capitalists, angel investors and other members of the business community. Each PEPITE provides

---

[2] As in the case of patents, this is likely an underestimation of the ASOs created each year (see Mustar, 2015). Based on a very broad definition of academic start-up firms, the French Ministry of Research provides an estimate of approximately 100 public research start-ups created every year (MESR, 2009). This number of ASOs created every year (between 55 and 100) has remained fairly stable for the last 15 years (Mustar, 2012).

advisory, support and training services for students and recent graduates who develop a start-up or a start-up project.

Students obtaining this status enjoy the following benefits:

- They have access to a dedicated training program for would-be entrepreneurs and thematic training modules like conferences, workshops and tailor-made programs.
- They can balance their studies and entrepreneurial activities. They have time allowance and flexibility for their studies. For example, they have the opportunity to replace their internship or final year project with their entrepreneurship project.
- They are assigned mentors: a two-member mentorship team, one from the faculty and one from the business world for each project leader.
- They have access to regional and national grant opportunities, such as the PEPITE Competition (a special type of business plan competition).
- They have access to tools that enable them to develop their venture: shared working spaces and "fablabs".
- They have the same healthcare coverage as students after they graduate (provided that he/she is enrolled in the Student Entrepreneur Degree).

The French government's program to confer the status of student entrepreneur was launched in September 2014. To achieve this status, students and young alumni (under 28 years of age) must submit an application to the PEPITE associated with their university or other type of higher educational institution. Then, after an oral presentation and a debate with the candidate, the selection committee expresses an opinion, favorable or unfavorable. If the opinion is favorable, the ministry of higher education awards the status to the student. The selection is made on the basis of the quality of the entrepreneurial project and the qualities and motivation of the candidate. Data collected from the 30 PEPITE in March 2018 show that 3,576 students submitted applications for this status for the academic year 2017–2018.

This support for students with an entrepreneurial project has been very successful since its establishment. According to PÉPITE France, the number of students obtaining entrepreneurial status has been monotonically increasing since 2014, rising from 637 in the 2014–2015 academic year to 3,576 students in the 2017–2018 academic year. This represents an increase of nearly 1,000 students per year. For the most recent academic year, more than 800 companies were registered by these students (including 200, as self-employed entrepreneurs).

These figures understate the enthusiasm for entrepreneurship among students, since not all students who wish to start a business apply for this status, especially those in engineering or management schools that already have procedures in place to encourage and support them. Nor do these figures include numerous young alumni who start a business after their first job.

Comparing these figures with the start-ups created by staff, it is evident that the number of student and young graduate start-ups per year is approximately 20 times greater. This ratio is similar to that of the U.K. noted earlier. Initial figures indicate that 70 percent of student entrepreneurs are students pursuing training and 30 percent are recent graduates. Almost three quarters (74 percent) are male. Two-thirds are aged between 20 and 24 years (Table 4). The average and median age is 23.

The 53 winning projects of the PEPITE 2015 Competition (known in France as the *Prix PEPITE de l'Entrepreneuriat Etudiant*) cover a surprisingly wide range of domains, with one-third in ICT followed by eco-technologies (Table 5).

Although this National Student Entrepreneur status was created in France, it is now being transferred with the support of French officials to other countries, including Belgium, Morocco, Tunisia and Lebanon. Students from these countries should gradually benefit from dual support

Table 4:   Students applying for "student entrepreneur" status in France

Age range	%
19 years old and under	4
20–24 years old	65
25–29 years old	27
30 years old and more	4
Total	100

Source: Prix PEPITES, December 2015.

Table 5: Industry distribution of the student start-ups in France (2015)

Industry	%
Information communication technologies	30
Environmental technologies	13
Unregistered	11
Computing	9
Biotechnology/Health	6
Security	6
Electronics	6
Food	4
Transport	2
Book market place	2
Education	2
Energy	2
Art	2
Mechanical	2
Building	2
Leisure	2
Total	100

Source: Prix PEPITES, December 2015.

(a teacher and an entrepreneur), the replacement of the final internship by the entrepreneurial project, financing, networking, access to a co-working space, etc.

Case study of a successful student venture from MINES ParisTech: Expliseat created by students during their studies in France

Expliseat[3] is a company created in March 2011 by three young engineers, while two of them were students at MINES ParisTech — Benjamin Saada,

---

[3]These pages are taken from the story of Expliseat in Mustar (2019).

the CEO, and Vincent Tejedor, and the third, Jean-Charles Samuelian, who was at the Ecole des Ponts Paristech. The company's objective was to produce an aircraft seat three times lighter than the seats then fitted to Airbus or Boeing aircrafts. At the time, the seats were made up of hundreds of parts assembled on an aluminum structure and weighed nearly 12 kilos. To limit this weight, the three founders designed an ultra-light seat made of a composite and titanium structure with only 30 parts. It weighed 4 kilos and was the lightest on the market. It allowed airlines to reduce the weight of their aircraft (by 2.2 tons on an A321), reduce their fuel costs (by $400,000 per year per aircraft) and reduce their CO_2 emissions (by 800 tons per year per aircraft).

At the beginning of their adventure, these young engineers knew nothing about aircraft seats or the aeronautics sector. In the early years, their project aroused skepticism and mockery. "At each step", Benjamin reminds us, "we heard the same argument: 'You'll never make it!' Your concept is great but you'll never be able to lighten the weight of the seat!'". When we did, we were told "It's great, but you'll never get it certified". When we achieved certification in Europe and the United States: "'You will never succeed in large-scale production of seats, especially not in France'. Yet, we have also succeeded in this, even though all economy class seats are produced in low-cost countries."

The path that Expliseat followed to successfully design, build, certify, produce and sell its seats was long and tortuous. It took the company 3 years to make its first sale to a Charter airline in its initial target market, that of the Boeing and Airbus B373 and A320 families. This market proved to be a dead end. Expliseat abandoned it, making a pivot to the regional aircraft market. It was only in 2016, 5 years after its creation, that the company took off when several regional companies and the manufacturer ATR placed orders with it.

Since then, it has grown by 250 percent per year. In 2018, it had a large cash position and generated several millions of euros in revenue and several tens of millions of euros of orders. The company now has a global presence in Canada, United States, Latin America, Central America with Nicaragua, Caribbean, Asia, Southeast Asia, New Caledonia and the Pacific. The year 2018 was also the year in which the major aircraft manufacturers (Airbus and Boeing) finally included Expliseat's seats in their catalogues. The company has 30 employees, almost all engineers, who design, sell and define the manufacturing methods for its seats. It does not have a factory but relies on numerous subcontractors, mainly French, to manufacture the various components and on an equipment manufacturer to carry out the final assembly. In these companies, nearly a hundred employees work full-time for Expliseat. Benjamin Saada maintains strong relationships with MINES ParisTech where he gives lectures to students on start-up creation, business management and materials science. He also advises students who have start-up projects. Notably, he has recruited several students from MINES ParisTech.

Case study of a successful student venture from MINES ParisTech: Criteo created by young alumni in France

Franck Le Ouay obtained his engineering degree from MINES ParisTech in 2000. His first job upon graduation was with Microsoft in Seattle and then in Aachen in Germany. In the fall of 2004, he resigned and returned to France to develop a universal product recommendation system based on the analysis of Internet users' behavior. A few months later, Romain Niccoli, a friend of Franck's who was in the same class at MINES ParisTech and who, like him, started his career at Microsoft in Seattle, joined him to develop the project then called Trustopia. In April 2005, they both joined the Parisian incubator, Agoranov, which had been created by universities and research organizations. Their project subsequently changed its name to become Criteo.

It was in the premises of Agoranov, in September 2005, that they met Jean-Baptiste Rudelle who came to present his project to an investment committee. This project turned out to be almost the same as that of Franck and Romain. Jean-Baptiste was 10 years older than them and had already created a successful venture capital-backed company specializing in customizing ringtones for mobile phones that had been sold to an American company.

Franck, Romain and Jean-Baptiste decided to join forces because their profiles were highly complementary. Franck and Romain have outstanding technological skills that enabled them to create the algorithm behind Criteo as well as managerial qualities gained from their time as project managers at Microsoft. Jean-Baptiste has experience in creating and developing a start-up, but also in fund-raising. Jean-Baptiste became CEO of Criteo, with Franck as the Chief Scientist and Romain as the CTO.

During the period 2005–2008, the company, financed by VC, sought to identify both its product and its market. These years involved experimentation during which the company pivoted several times. In 2008, it found its product and business model. Its product is a form of display advertising. Criteo works with Internet retailers to serve personalized online display adverts to consumers who have previously visited the advertiser's website. Its solution operates on a pay per click/cost per click (CPC) basis. This means that advertisers pay only for the consumers that click on the banner and return to their website.

Criteo has experienced very strong growth since 2009. In 2010, it opened an office in Silicon Valley to enhance its profile in the U.S. In 2013, Criteo entered NASDAQ in New York with a capitalization of $2 billion. Since then, its growth has been continuous: from 800 employees in 2013 to 3,000 in 2018. Criteo currently operates in a total of 30 markets around the world and is headquartered in Paris.

Franck and Romain have kept strong links with MINES ParisTech. They regularly return to present talks on the progress of their company to students interested in entrepreneurship. They also serve as coaches or mentors to many projects and start-ups from MINES ParisTech. The company currently sponsors the MINES ParisTech Criteo Entrepreneurship Award.

Italy

Fini *et al.* (2016a, b) provide recent evidence from Italy on student entrepreneurship, based on the population of students graduating from 64 Italian universities in 2014. Of 61,115 undergraduate (bachelor) and graduate (master) students, who graduated between September and December 2014, 1,664 (2.7 percent) were student entrepreneurs (i.e., students who had created a new venture during their university study or before starting university), and 2,232 (3.8 percent) were nascent entrepreneurs (i.e., students who were engaged in some entrepreneurial activities).

Among those creating a venture, 66 percent did so during their university studies, whereas 34 percent established a venture before attending the university.

The Italian study provides interesting insights regarding the academic fields chosen by student entrepreneurs (see Table 6). Rather than being dominated by students in STEMM subjects (i.e., Science, Technology, Engineering, Math and Medicine), as might be expected, it appears that more than about half of the student entrepreneurs (including both actual and nascent or would-be entrepreneurs) completed a degree in the social sciences (i.e., Economics-Statistics, Education, Law, Linguistics, Political Science, Sociology and Psychology), and about 40 percent in STEMM

Table 6:   Areas of study of student entrepreneurs in Italy

Variable		Entrepreneurs (*n* = 1,664)		Nascent entrepreneurs (*n* = 2,232)		Non-entrepreneurs (*n* = 57,219)	
		n	%	*n*	%	*n*	%
Field of Study	Agriculture and Veterinary	62	3.7	87	3.9	1,435	2.5
	Architecture	86	5.2	106	4.7	2,135	3.7
	Chemistry–Pharmaceutical	40	2.4	66	3.0	2,030	3.5
	Economics–Statistics	300	18.0	491	22.0	7,924	13.8
	Engineering	179	10.8	304	13.6	7,276	12.7
	Education	82	4.9	78	3.5	2,359	4.1
	Geology–Biology	42	2.5	85	3.8	3,000	5.2
	Law	123	7.4	108	4.8	2,855	5.0
	Linguistics	40	2.4	62	2.8	2,899	5.1
	Literature	114	6.9	149	6.7	3,862	6.7
	Medicine	255	15.3	241	10.8	11,276	19.7
	Physical Education	38	2.3	51	2.3	1,177	2.1
	Political–Social	189	11.4	243	10.9	4,701	8.2
	Psychology	70	4.2	92	4.1	2,421	4.2
	Scientific	42	2.5	68	3.1	1,843	3.2

Source: Fini *et al.* (2016a, b).

disciplines, with the remainder studying humanities or physical education.

But, as we saw earlier with the experience of Alex Kantos in the U.K., engaging in entrepreneurship requires deep commitment which if not well managed can have adverse implications on a student's academic performance. Fini and colleagues' study shows that for Italy, students who engage in entrepreneurship are late in completing their degrees compared to non-entrepreneurs and their performance is average. There may therefore be serious opportunity costs for student entrepreneurship, especially if venture efforts are not successful or students do not receive program support required to help them succeed. Students need to be aware of and take account of these opportunity costs rather than getting swept up in the hype about starting a venture.

There is a clear age difference between students who become entrepreneurs in Italy and those who do not. The average age of student entrepreneurs was 30 years, compared to 27 years for nascent entrepreneurs and 25.5 years for non-entrepreneurs. Around 40 percent of student entrepreneurs are women. About 60 percent of student entrepreneurs reside in the region where they graduated, a higher proportion than for nascent and non-entrepreneurs. The authors also report a finding that relates to the North/South divide in Italy. They suggest that student entrepreneurship in southern regions of Italy appears to be substituting for the lower number of job opportunities outside the more industrialized, developed northern regions of Italy. This finding has important implications for university administrators who assert that student entrepreneurship can contribute to regional economic development, especially in towns and cities that are experiencing sluggish job growth.

Over 75 percent of student and nascent entrepreneurs were induced to become entrepreneurs by family, as well as fellow students. While university professors were reported to provide the most entrepreneurial competences to help student and nascent entrepreneurs, almost 80 percent of both categories indicated that there were no entrepreneurship courses in their universities' programs. The nascent student entrepreneurs attached greater importance than the actual student entrepreneurs to there being a need for an entrepreneurship course. Only 11.5 percent of student entrepreneurs and 17.2 percent of nascent entrepreneurs enrolled in an entrepreneurship course. The incidence of participation in business plan competitions was similarly low. Only 7 percent of student entrepreneurs and 9.6 percent of nascent entrepreneurs participated in a business plan competition. This may in no small part be due to the patchy provision of educational support for entrepreneurship in the country. Indeed, Fini *et al.* (2016a, b) recommend that there is a need in Italy for more systematic curricula programs aimed at developing entrepreneurial skills among students.

Such support might help students in starting more viable ventures. According to the study, 37 percent of student entrepreneurs in Italy were no longer involved with their ventures either because they had exited the company or because the venture was no longer active. About 43 percent of these non-active entrepreneurs indicated that the most important reason for closing the business was that revenues/profit were below expectations.

Case study of a successful student venture in Italy: Fatture in cloud

Fatture in Cloud, a highly innovative firm in online invoicing, was a start-up developed by Daniele Ratti in 2013, while he was a software engineering student at the University of Bergamo in Italy. The software program Daniele created was used to develop a new niche in the IT services market. He launched the new company in 2014, which was based on this program. Specifically, Fatture in Cloud provided online invoicing services to SMEs and self-employed individuals.

The firm was successful because it effectively disrupted the invoicing market in Italy, offering customers a new way to handle their billing and invoicing needs. Users were able to use a service that was not only reliable, but also delivered features that traditional invoicing companies were not providing at the time, such as the ability to exchange information in real time. More importantly, all data and documents are accessible by customers 24/7 on any computer or smartphone. In 2015, Ratti sold a 51 percent stake in Fatture in Cloud to the American market leader in accountancy software, TeamSystem, for more than €1 million. Fatture in Cloud currently has over 20,000 active users and generates an annual revenue of €2 million ($2.3 million).

Africa

Africa has numerous emerging economies that are supporting student entrepreneurship. With nearly 1.3 billion inhabitants, it is experiencing unprecedented economic and social transformation. In 2018, according to the World Bank, 6 of the 10 fastest growing countries in the world were African. The economic growth rate is expected to exceed 3.2 percent in 2018, with a figure of more than 6 percent for about 10 countries. This growth is driven by the agricultural and infrastructure sectors. It is also linked to strong population growth, with an urban population that is expected to double in the next 25 years. The African population is quite young: 65 percent of Africans are under 25 years of age.

Driven by this growth, start-up ecosystems are developing in the continent.[4] One of the factors behind their success is the mobile phone penetration rate, which exceeds 80 percent. While Africa has more than 400 Tech Hubs, the emergence of start-ups remains very heterogeneous across regions and countries. This appears to be most dynamic in English-speaking Africa. For example, Kenya, with more than 1,500 technology start-ups and a mobile phone penetration rate of 88 percent in 2016, is one of Africa's leaders in the digital economy. Innovation centers, incubators, accelerators, co-working spaces and investment funds are multiplying, thanks to a proactive public policy in the field of entrepreneurship.

Contributing to this environment, universities play an important role in developing an entrepreneurial spirit among their students. They support them in their start-up creation project, in particular through incubators. The most famous are the Chandaria Business and Incubation Center at Kenyatta University and the iLab Africa at Strathmore University. Chandaria Business Innovation and Incubation Centre (Chandaria-BIIC) was launched in July 2011 to help students and other Kenyan entrepreneurs in need of support.[5] In line with Kenya's Vision 2030 and the Kenyatta University's current Strategic and Vision Plan, Chandaria-BIIC focuses on supporting up to 120 start-ups per year (70 percent Kenyatta

---

[4]The data in this section was provided by Rania Haira, whom we would like to thank.
[5]Information about Chandaria-BIIC comes from its website http://www.ku.ac.ke/chandaria-biic/.

University students and 30 percent Non-KU). According to the university's website, it aims to blend applied research with innovation and the establishment of start-ups as well as predispose Kenyatta University students and Kenyans in general toward being job creators rather than job seekers.

@iLabAfrica is a Centre of Excellence in ICT innovation and development based at Strathmore University. The research center is involved in interdisciplinary research, student engagement, collaboration with government, industry and other funding agencies. The goal of the center is "to provide an environment that promotes technological innovation and business support structures and policy direction to harness the potential of ICT as a genuine tool for sustainable development".

@iBizAfrica is the business incubator which carries out the Entrepreneurship and Incubation theme of @iLabAfrica. This organization acts as a classical incubator, fostering an environment for entrepreneurs to develop, nurture and exchange their innovative ideas, providing mentoring and services to start-up companies such as seed capital, legal advice, financial expertise, relevant training and physical resources, providing business incubation facilities and establish linkages with other incubation centers in the country and around the world, forging industry partnerships, and acting as a focal point for investors to engage with potential technology entrepreneurs. The website indicates that there have been more than 500 events and more than 300 start-ups have been supported by 200 mentors.

A key problem for many African universities is that many students have difficulty finding employment after graduation. In Nigeria, it has been estimated that nearly a quarter of graduates are unemployed when they leave university. In Kenya, two-thirds of students expect that they will be self-employed when they graduate.[6] In Tanzania, which had a growth rate of 6.8 percent in 2018, many students become entrepreneurs while they are still studying. In Tanzania, many initiatives exist to foster the creation of new ventures by students or recent graduates, notably incubation hubs and business plan competitions. For example, the Tanzania

---

[6] https://www.bbc.com/news/av/business-37727760/student-start-ups-are-next-big-thing-in-tanzania.

Universities Entrepreneurship Challenge is an initiative aimed at getting more and more university students interested in choosing entrepreneurship as their career path so that they become job creators rather than job seekers.

Student Entrepreneurship and Entrepreneurship Education

Evidence from leading research universities, such as MIT (Roberts and Eesley, 2011) and Stanford (Lazear, 2005), show a greater incidence of venture creation by students than by faculty. MIT professors Hsu *et al.* (2007) find that students are also getting younger when they start their venture. The entrepreneurial spirit has also invaded the top business schools. A Harvard Business School study showed that approximately 5 percent of alumni create a new venture within 1 year after graduation (Lerner and Malmendier, 2013). More broadly, across the U.S., recent science and engineering graduates are twice as likely as faculty members to start a business within 3 years of graduation (Astebro *et al.*, 2012). Moreover, these graduate start-ups were found to be of high quality. This evidence appears to call into question strategies by universities that focus on increasing start-ups led by faculty and suggests a need to at least complement them by developing systematic mechanisms to support start-ups by both alumni and current students.

Some graduates create a venture upon graduation or they may seek to gain industrial and commercial experience first, before becoming entrepreneurs. Global evidence suggests that students tend to have a preference for obtaining commercial experience through employment upon graduation before becoming entrepreneurs (Sieger *et al.*, 2011). Gaining industrial and commercial experience first may contribute to the entrepreneurial venture that is created being more viable. Some support for this conjecture is hinted at in a study of former employees of universities in Sweden who became entrepreneurs either directly after leaving academia or indirectly after gaining work experience outside the university (Wennberg *et al.*, 2011). This study shows that enterprises created by university graduates who had business experience were more likely to grow and survive than those created by individuals who launched ventures upon graduation.

An important issue concerns the factors that influence students' nascent entrepreneurial behavior, that is, their intentions and initial preparations to start a venture, as well as their actual creation of a venture. There is, at best, mixed evidence concerning the relationship between years of schooling and entrepreneurial entry (van der Sluis *et al.*, 2008). However, other individual characteristics appear to play an important role. Students with a more balanced portfolio of human and social capital appear more willing than those with more specialist human and social capital to become entrepreneurs, according to German evidence (Backes-Gellner and Moog, 2007). Additional evidence suggests that gender matters, in terms of the impact of entrepreneurship education (Wilson *et al.*, 2007). For example, a study of business students and professors from 25 academic departments showed that entrepreneurship education and industry ties are related to self-employment intentions only for males (Walter *et al.*, 2013).

Evidence from business and economics students at European universities shows that while individual characteristics such as age, gender and parental self-employment are the most important drivers, contextual factors are also important (Bergman *et al.*, 2016). Nascent entrepreneurial actions appear to be influenced by organizational characteristics, such as the prevalence of fellow students who have attended entrepreneurship education. However, these factors do not seem to support the actual establishment of a new venture. Instead, it appears that the creation of a new venture is less dependent on the university context than on regional characteristics (the region where the university is located). Evidence from Italy and Sweden indicates that students are more likely to start their ventures in the region where the student entrepreneurs complete their studies, even after controlling for birth region (Baltzopoulos and Brostrom, 2013; Colombo *et al.*, 2015; Fini *et al.*, 2016b). This indicates that elements of the entrepreneurial ecosystem have an important effect on the extent and nature of student entrepreneurship, a subject we return to in Chapter 3.

Conclusion: The Need for a Student Entrepreneurship Ecosystem

Given the evidence we have presented on growth in student entrepreneurship, we need to understand the ecosystem that enables start-ups to emerge and develop. Entrepreneurial ecosystems regulate the nature and quality of entrepreneurial activities by shaping the direction and potential rewards associated with opportunity identification, creation, and pursuit, and by also establishing the types of organizational forms that will be accepted as legitimate (e.g., the creation of a firm). They involve multi-level processes and stakeholders, multiple actors and multiple contexts (Isenberg, 2010). Developing an ecosystem framework provides the basis for more effective interventions at both government and university levels, especially as we have seen that the number of ventures created by students and graduates far outweighs those created by faculty (Astebro *et al.*, 2012).

An entrepreneurial ecosystem for students has many dimensions as shown in Figure 1 which we use as an organizing framework for the

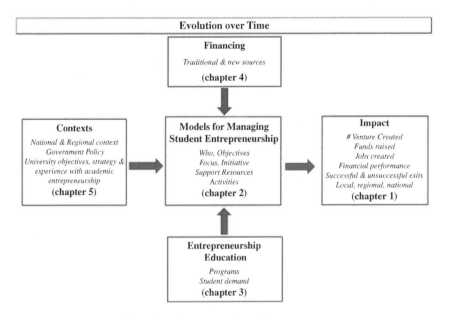

Figure 1:   Ecosystem for student start-ups

chapters in this book. At its core it comprises different models for managing student entrepreneurship activities at universities. Key elements of these models relate to the actors involved in their initiation, their focus and objectives, their support resources and their activities. In Chapter 2, we develop a typology of these different models.

The student entrepreneurship ecosystem also includes entrepreneurship courses and programs (including degree programs and certificate programs). Reflecting an increase in student demand, there has been a recent proliferation of initiatives that move beyond traditional class room teaching and business plan competitions have taken place in the absence of an overall framework for understanding the ecosystem that supports student entrepreneurship. Many of these initiatives involve experiential learning, including garages, creativity and ingenuity labs, and other programs designed to ensure that students actually create a venture. For example, Nottingham University in the U.K. has created an Ingenuity Centre worth £5.2 million which provides diverse working and meeting spaces, with 40 hot-desks, 10 office pods, 4 meeting and lounge spaces including 2 state-of-the-art white boardrooms and a lounge area where members can relax and reflect. We explore the developments in entrepreneurial education at universities and their implications for student entrepreneurship in Chapter 3.

Also feeding into the emergence and development of student ventures are various sources of funding. It is well recognized that entrepreneurial finance sources need to involve the provision of both funds and also mentoring support. While some traditional funding sources for entrepreneurs are generally inappropriate for students' ventures, new funding sources are emerging specifically targeted at students. We discuss the various funding sources available for student entrepreneurship in Chapter 4.

Such an ecosystem develops in the context of the vital formal and informal rules and regulations governing a national and regional context (North, 1990; Autio *et al.*, 2014). Formal institutional features include the rule of law and property rights. Informal institutional mechanisms refer to sanctions, traditions and codes of conduct. These institutions constitute the rules of the game. For entrepreneurship, they can also include norms, such as respect for entrepreneurship as an honorable activity, a credible field of study and area of research and tolerance of failure. More tangible support can also come from public infrastructure to support entrepreneurship, such as incubators/accelerators (Pauwels *et al.*, 2016), as well as grants and successful entrepreneurs serving as mentors. What Kenney and Patton (2011) called "entrepreneurial support networks" (e.g., actors as venture capitalists, lawyers and accountants) are also formal institutions assisting in the formation and growth of entrepreneurial firms. Informal institutions include the wider culture (Stephan and Uhlander, 2010) and social norms (Webb *et al.*, 2009).

The ecosystem for student entrepreneurship also develops in the context of particular universities in terms of their aims and objectives, their strategies and their experiences with academic entrepreneurship. We discuss these contextual factors in Chapter 5.

In Chapter 6, the final chapter, we draw together the insights developed throughout the book to identify implications for university and governmental policies to develop the student entrepreneurship ecosystem, as well as set out directions for further research.

References

Astebro, T., Bazzazian, N., and Braguinsky, S. 2012. Startups by recent university graduates and their faculty: Implications for university entrepreneurship policy. *Research Policy*, 41(4), 663–677.

Autio, E., Kenney, M., Mustar, P., Siegel, D., and Wright, M. 2014. Entrepreneurial innovation: The importance of context. *Research Policy*, 43, 1097–1108.

Backes-Gellner, U., and Moog, P. 2007. Who chooses to become an entrepreneur? The jacks-of-all-trades in social and human capital. *Journal of Socio-Economics*, 47, 55–72.

Baltzopoulos, A., and Broström, A. 2013. Attractors of entrepreneurial activity: Universities, regions and alumni entrepreneurs. *Regional Studies*, 47(6), 934–949.

Bergman, H., Hundt, C., and Sternberg, R. 2016. What makes student entrepreneurs? On the relevance (and irrelevance) of the university and the regional context for student start-ups. *Small Business Economics*, doi: 10.1007/s11187-016-9700-6.

Boh, W. F., De-Haan, U., and Strom, R. 2016. University technology transfer through entrepreneurship: Faculty and students in spinoffs. *The Journal of Technology Transfer*, 41(4), 661–669.

Brin, S., and Page, L. 1998. The anatomy of a large-scale hypertextual web search engine, In: *Proceedings of the 7th International World Wide Web Conference*, April 14–18, Brisbane, Australia, pp. 107–117.

Colombo M., Piva E., and Rossi-Lamastra C. 2015. Student entrepreneurs from technology-based universities: The impact of course curriculum on entrepreneurial entry. *Imperial Innovation and Entrepreneurship Conference*, June 18–19, 2015, Royal Society of London.

Council for Science and Technology. 2016. Strengthening entrepreneurship education to boost growth, jobs and productivity.

Eesley, C., and Miller, W. 2012. Stanford University's Economic Impact via Innovation and Entrepreneurship. Stanford University, October 2012.

Fini, R. Fu, K., Rasmussen, E., Mathison, M., and Wright, M. 2016a. Determinants of university startup quantity and quality in Italy, Norway and the U.K. ERC Working Paper.

Fini R., Meoli A., Sobrero M., Ghiselli S., and Ferrante F. 2016b. *Student Entrepreneurship: Demographics, Competences and Obstacles.* Consorzio Interuniversitario AlmaLaurea.

Hart, D. 2004. On the Origins of Google (Washington, DC: NSF: August 17, 2004), https://www.nsf.gov/discoveries/disc_summ.jsp?cntn_id=100660.

Hart, M., and Bonner, K. 2015. GEM U.K.: Graduate Entrepreneurship. Aston Business School and Enterprise Research Centre.

HESA 2015. The Higher Education — Business and Community Interaction (HE-BCI) Survey 2013/4. HEFC.

HESA 2017. The Higher Education — Business and Community Interaction (HE-BCI) Survey 2015/6. HEFC.

Holstein, J., Starkey, K., and Wright, M. 2018. Strategy and narrative in higher education. *Strategic Organization*, 16, 61–91.

Hsu, D. H., Roberts, E. B., and Eesley, C. E. 2007. Entrepreneurs from technology-based universities: Evidence from MIT. *Research Policy*, 36(5), 768–788.

Isenberg, D. 2010. The big idea: How to start an entrepreneurial revolution. *Harvard Business Review*, 88, 40–50.

Kenney, M., and D. Patton. 2011. Does inventor ownership encourage university research derived entrepreneurship? A six university comparison. *Research Policy*, 40(8), 1100–1112.

Lazear, E. P. 2005. Entrepreneurship. *Journal of Labor Economics*, 23, 649–680.

Lerner, J., and Malmendier, U. 2013. With a little help from my (random) friends: Success and failure in post-business school entrepreneurship. *Review of Financial Studies*, 26(10), 2411–2452.

MESR 2009. Recherche et Développement, Innovation et Partenariats 2008, Ministère de l'Enseignement Supérieur et de la Recherche, Direction Générale pour la Recherche et l'Innovation, Septembre, p. 87.

Mustar, P., Renault, M., Colombo M. G., Piva, E., Fontes, M., Lockett A., Wright, M., Clarysse, B., and Moray, N. 2006. Conceptualising the heterogeneity of research-based start-ups: A multi-dimensional taxonomy. *Research Policy*, 35, 289–308.

Mustar, P. 2012. The French Experience: for a broader view of academic entrepreneurship, paper presented at the session: assessing academic entrepreneurship: a comparative analysis of Europe vs. US, Academy of Management Annual Conference, 2012, Boston, 2–8 August.

Mustar, P. 2015. "Tools, rationale and economic impact of the French public policy in the 2000s to foster the creation of academic spin-off firms", paper presented at the International Science-Based Entrepreneurship Workshop, May 7–8, 2015, University of Bologna Business School, Italy.

Mustar, P. 2019. Process of creation of start-ups by students and young graduates of MINES ParisTech, project in progress.

Nabi, G., Liñán, F., Fayolle, A., Krueger, N., and Walmsley, A. 2017. The impact of entrepreneurship education in higher education: A systematic review and research agenda. *Academy of Management Learning and Education*, 16(2), 277–299.

North, D. 1990. *Institutions, Institutional Change and Economic Performance*. Cambridge: Cambridge University Press.

OECD 2015a. Entrepreneurship at a glance 2015. OECD Publishing, Paris. Pitchbook Universities Report 2016–2017.

Pauwels, C., Clarysse, B., Wright, M., and Van Hove, J. 2016. Understanding a new generation incubation model: The accelerator. *Technovation*, 50–51, 13–24.

Roberts, E. B., and Eesley, C. E. 2011. Entrepreneurial impact: The Role of MIT. *Foundations and Trends in Entrepreneurship*, 7(1), 1–149.

Senor, D., and Singer, S. 2009. *Start-up nation: The story of Israel's economic miracle*. New York: Grand Central Publishing.

Siegel, D. S., and Wessner, C. 2012. Universities and the success of entrepreneurial ventures: Evidence from the small business innovation research program. *Journal of Technology Transfer*, 37(4), 404–415.

Sieger, P., Fueglistaller, U., and Zellweger, T. 2011. Entrepreneurial Intentions and activities of students across the world. International report of the GUESSS Project 2011. St. Gallen.

Singer, P. L. 2014. Federally Supported Innovations: 22 Examples of Major Technology Advances That Stem From Federal Research Support, mimeo, The Information Technology and Innovation Foundation, Washington, D.C.

Stephan, P. 2009. Tracking the placement of students as a measure of technology transfer. In: Gary D. L. (ed.), *Measuring the Social Value of Innovation: A Link in the University Technology Transfer and Entrepreneurship Equation.* Advances in the Study of Entrepreneurship, Innovation & Economic Growth, Volume 19, Emerald Group Publishing Limited, pp. 113–140.

Stephan, U., and Uhlaner, L. M. 2010. Performance-based vs. socially supportive culture: A cross-national study of descriptive norms and entrepreneurship. *Journal of International Business Studies*, 41, 1347–1364.

University of York 2014. University of York student awarded Duke of York Young Entrepreneur Award. https://www.york.ac.U.K./news-and-events/news/2014/dU.K.e-of-york/ [accessed October 8, 2018].

Van der Sluis, J., Van Praag, M., and Vijverberg, W. 2008. Education and entrepreneurship selection and performance: A review of the empirical literature. *Journal of Economic Surveys*, 22(5), 795–841.

Walter, S. G., Parboteeah, K. P., and Walter, A. 2013. University departments and self-employment intentions of business students: A cross-level analysis. *Entrepreneurship Theory and Practice*, 37(2), 175–200.

Webb, J., Ireland, D., Tihanyi, L., and Sirmon, D. 2009. You say illegal, i say legitimate: Entrepreneurship in the informal economy. *Academy of Management Review*, 34(3), 492–510.

Wennberg, K., Wiklund, J., and Wright, M., 2011. The effectiveness of university knowledge spillovers: Performance differences between university spinoffs and corporate spinoffs. *Research Policy*, 40, 1128–1143.

Wilson, F., Kickul, J., and Marlino, D. 2007. Gender, entrepreneurial self-efficacy, and entrepreneurial career intentions: Implications for entrepreneurship education. *Entrepreneurship Theory and Practice*, 31(3), 387–406.

Chapter 2

Models for Managing Student Entrepreneurial Activities

Introduction

The ecosystem outlined in Chapter 1 highlighted the influence of several factors on incubator activities relating to student entrepreneurship. In recent decades, many universities have developed an ecosystem that promotes student entrepreneurship. These institutions have offered entrepreneurship courses for many years. More recently, however, they are also allocating space and different kinds of resources to student entrepreneurs to allow them to develop their start-up or to make prototypes of their future products. In this chapter, we expand on the element in Figure 1 of Chapter 1 relating to the models for managing student entrepreneurship by developing a framework for understanding the variety of configurations in which student entrepreneurship is generated.

Based on our detailed case studies, selected from a range of universities from a dozen countries, we conclude that there is substantial heterogeneity in university efforts to promote student entrepreneurship. We selected both universities with a full range of disciplines as well as specialist technical universities. The case studies involved face-to-face interviews with university faculty directly involved in designing and leading the approaches, administrators and student entrepreneurs using the facilities and programs. In addition, we gathered archival data from the universities' websites and hard copy documentation, as well as secondary data relating to specific detailed case studies. This approach enabled us to triangulate different perspectives.

Specifically, we rely on case evidence from Imperial College London, Cambridge University, University College London, Cass Business School, the University of Nottingham and the Royal College of Art from the U.K. It also includes the experience of Continental European universities, engineering schools and business schools in France (MINES ParisTech, Paris Sciences Lettres — PSL Research University); in some Nordic countries, such as Aalto University in Finland; and in Southern European countries, notably IQS in Spain and the University of Bologna in Italy (Sala and Sobrero, 2018). In addition, we used evidence from the U.S., notably from the University of Albany, SUNY and MIT, and from the University of Adelaide in Australia.

To guide our data collection and analysis, we adapt the framework utilized by Clarysse *et al.* (2005) in analyzing higher education and research institutes' incubation of academic spin-offs. Clarysse *et al.*'s framework distinguishes between the resources and activities devoted to incubation and identifies differences in the objectives of the organizations and in the kinds of ventures being created. We also identify the source of the initiative for student entrepreneurship. While Clarysse *et al.* (2005) only examine instances when a university or research institute had an active spin-off strategy, we adopt a broader view, given that student entrepreneurship is a more emergent phenomenon. As such, we also extend the approach of Boh *et al.* (2016) based on technology commercialization, who built their framework on an examination of eight leading research universities in the U.S. We nuance their distinction between structured and *laissez-faire* approaches to better highlight the focus of the initiative and also their distinction between internal and external resources, in order to focus on the strength and nature of resource support by actors internal and external to the university.

We find that some cases have resulted from bottom-up initiatives, while others have been based on top-down policies. Some universities have mobilized substantial resources to devote to these activities, both from inside and outside the institution, while others manage them on a modest budget. In some universities, these initiatives are centrally managed, while in others, they are implemented by one or two professors. Another model is when such activities are student-run.

As an attempt to make sense of this proliferation of initiatives and situations, we identify a typology of clear and distinct models and analyze these configurations.

A Typology of Student Entrepreneurship Models

Based on our fieldwork, we distinguish four different "ideal" types that enable us to compare the organizational/institutional configurations within universities that facilitate entrepreneurial activities and student start-ups.

A central question guides our approach and our typology: What is the driving force for entrepreneurial activities in universities and who are the key players and institutions involved? This approach is summarized in Figure 1 and Table 1 where we distinguish the existence or non-existence of a university strategy for student entrepreneurship (in terms of the recognition of its importance) on the *Y*-axis and the existence or non-existence of university support for student entrepreneurship (in terms of devoted resources) on the *X*-axis. In our simplified model, the university

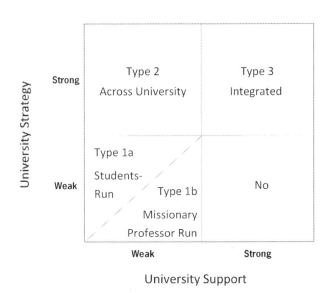

Figure 1:   Typology of policies relating to student entrepreneurship

Table 1:   Characteristics of the four models for managing student entrepreneurship at universities

	Who	Objectives	Focus	Initiative	Resources	Activities	Examples
Student-run	Student entrepreneurship society; student club	Activities to support student entrepreneurial project	Narrow: student start-ups, creation of companies by students	Bottom-up	Limited. Based on voluntary service (some exceptions)	Business plan competition. Conferences	Aalto Entrepreneurship Society
Missionary professor-run	Business schools. science departments, individual faculty members	Support mechanisms beyond the traditional classroom offerings	Narrow: student start-ups	Bottom-up (benevolent look of the university management)	Limited: 1 or 2 missionary professors. A small space Low financial support	Broad range: ENT EDUC, competition, events, conferences	Entrepreneurship Unit MINES ParisTech
Across university	A light structure in large or new university, without a long entrepreneurial tradition	To support and develop student entrepreneurship	Narrow: students and recent graduates	Top-down by the management of the university	Very small team with modest space and a low financial support	Mentorship, training programs, access to co-working space	Imperial College Create Lab
Integrated	A powerful center for entrepreneurship (across all schools and disciplines)	To provide an integrated set of support mechanisms	To educate students in entrepreneurship and to support their start-ups	Top-down	High level of resources: strong team, different spaces, funds from university, rich alumni and corporate	Outreach, programs, events, infrastructure and state-of-the-art facility, academics	Martin Trust Center for MIT Entrepreneurship

strategy can be minimal, non-existent (−) or pivotal (+), while university support for student entrepreneurship can be low (−) or high (+).

Quadrant 1

With a weak or non-existing university strategy or support, the student entrepreneurship movement in some universities is being driven by some students or professors taking the initiative to lead activities and policy.

Within this quadrant, we distinguish Type 1a (Student-Run Entrepreneurship Policy) and Type 1b (Missionary Professor(s) Level-Run Student Entrepreneurship Policy). Resources are not available from the university, but it is possible for Types 1 and 2 to obtain funding from alumni or diverse sources.

Quadrant 2

This situation occurs when there is a willingness by a university to develop student entrepreneurship, but it allocates only a small amount of human and financial resources to fulfill this aim. We refer to approaches in this quadrant as Across University Intermediation for Student Entrepreneurship Policy.

Quadrant 3

This quadrant involves both a robust strategy and substantial resources from the university devoted to student entrepreneurship, creating a third type of initiative that we refer to as Integrated Management of Student Entrepreneurship Policy.

Quadrant 4

This quadrant is empty because if there is no desire on the part of the university to foster student entrepreneurship, it is logical to assume that there are no university resources deployed toward these activities. It is difficult to imagine a university that does not have a policy to develop student entrepreneurship but at the same time devoting resources to these activities! Of course, there may be universities with an aim to develop student entrepreneurship but no strategy, with resources being used "below the radar" to support student entrepreneurship.

In the following section, we elaborate on the elements of each model and for illustrative purposes present details relating to a stereotypical example of each type.

Type 1a: Student-Run Entrepreneurship Policy

With a non-existent explicit university strategy and support for entrepreneurship, students have taken the initiative to support entrepreneurship. They have created student entrepreneurship societies, student clubs or student associations as driving forces for entrepreneurial activities.

These societies, clubs or associations (hereinafter referred to as "clubs") play a crucial role in developing many activities to support entrepreneurial projects and the entrepreneurial spirit. These clubs are bottom-up initiatives. Some were created about 20 years ago (e.g., Cambridge University Entrepreneur was established in 1999). Typically, they are focused on the creation of companies by students. Clubs play an important role in the support of student entrepreneurship by organizing many networking events to create links between students from different departments or sometimes different universities.

Oftentimes run by students themselves, these clubs have traditionally organized visits to entrepreneurial firms and talks by well-known and successful entrepreneurs. Their main activity is to organize conferences about entrepreneurship as well as business plan competitions for which they find sponsors. Increasingly, clubs are developing their capabilities to enable more experiential learning regarding opportunity recognition and exploitation, the establishment of new ventures and seeking of funds.

Often these student clubs have limited resources. They are mainly based on voluntary service by students, along with alumni engagement, through their time and donations. They also attract sponsorship from firms. In some instances, these societies or clubs have substantial resources. This is the case with the Aalto Entrepreneurship Society (AaltoES), a powerful association created by students to develop entrepreneurship in this leading university in Finland.

By presenting in detail about AaltoES in Helsinki, we show what can be achieved when policy and entrepreneurial activities are fully organized and managed by students.

The Aalto Entrepreneurship Society[1]

In Finland, Aalto University was established on January 1, 2010 by merging the Helsinki School of Economics, Helsinki University of Technology and the University of Art and Design. In 2008, a small group of students in Helsinki were told by their professor to avoid entrepreneurship for their future professional career. Returning from a trip to Stanford, these students were astonished to hear this and realized that entrepreneurship was widely discouraged in Finland. In reaction to this, the students launched AaltoES to foster entrepreneurship at their university. In 2009, they organized their first event: a talk with the CEO of MySQL. Without real advertising, around 150 students came, showing a strong interest in entrepreneurship. The number of events they organized grew fast: about 50 per year mostly in a 1,500-square meter old industrial building given by the university. Since then, AaltoES has multiplied its initiatives in this field. Here are three of them: Startup Life, Startup Sauna and Slush.

- *Startup Life* is an internship program that sends engineering, design and business students from Finnish universities to work at start-ups based in Silicon Valley and San Francisco. It gives students the opportunity to join a start-up in a foreign entrepreneurship ecosystem as an employee for a duration of at least 3 months. At the beginning of this initiative, to find start-up placements for this program, a team of students went around Silicon Valley and pitched the concept. Now Startup Life has a good reputation, and the network created between California and Finland is self-sufficient. Startup Life is now expanding to Asia also: to Tokyo and Shanghai. "We offer Finnish university

---

[1] The data from Aalto University and AaltoES are from the review written by five students of MINES ParisTech (Paul Belin, Valentin Bourrelier, Claude Dao, Hubert Furno and Tom Monnier), under the leadership of Professor Philippe Mustar, after their study trip to Helsinki in February 2017.

students a hassle-free path to start-up internships in Silicon Valley. We give the students a chance for rocketing professional growth, for building international networks and for gathering unique, international working experience. And all this we do non-profit — for students, by students", *Aalto University, Startup Life internship program presentation.*[2]

- *Startup Sauna* is an accelerator for top-notch tech teams involving a 7-week program focused on finding the right product–market fit and go-to-market strategy. Startup Sauna is not about developing a product on a tech aspect, but focuses on business model development: how to pitch, improve marketing, obtain loans, recruit, etc. The program is free and does not take equity (as the majority of accelerators do). The selected projects are supervised by 70 volunteer coaches who have a first-hand experience on founding companies (serial entrepreneurs, angel investors, venture capital firms, etc.).

  In 2016, Startup Sauna had 1,500 start-ups applying. There are two sets of applicants each year: one in spring and one during the fall. Since its creation in 2010, 14 batches have produced 208 start-ups who raised more than $100 million. Startup Sauna is now the leading accelerator in the Nordics, Baltics, Eastern Europe and Russia.

- *Slush* is an AaltoES organizes more than 100 events per year (hackathons, pitching events, networking events, etc.). The most famous and important is Slush.

  Slush is a 2-day international start-up and investor event, created in 2008 and organized annually in Helsinki. Commitment needs to be high: it takes place during the coldest season in Finland! However Slush has grown and is now the biggest event of this kind in Europe. In November 2016, Slush gathered 17,000 attendees — including 2,300 start-ups, 1,100 investors and 600 journalists from 100 countries. It is organized by 2,000 volunteer students, mostly from Aalto University. Some of its major sponsors are Google, PwC, Samsung, Microsoft and the Finnish success story — Supercell. From a 300-person assembly, this event has grown to become a world-renowned event.

---

[2] https://into.aalto.fi/pages/viewpage.action?pageId=19638752.

These activities are complemented by the Design Factory, which has created a community of researchers, students and staff from different schools of Aalto University, entrepreneurs and company representatives. At the creation of this "new" university, the first official building was the Aalto Design Factory. This was set up as an experimental project open to the whole university and offering students infrastructure (3,000 square meters of working rooms, laboratories, prototyping rooms, coffee rooms, etc.), machines (three-dimensional printers, milling machines, grindstones, etc.), equipment (electrical and mechanical equipment, working tools, etc.) and human resources (machine operators to accomplish tasks such as manipulation of the machines for and with the students) to develop their projects. One of the main activities of the Design Factory is the Product Development project, a student-led program in cooperation with existing companies lasting 1 year, as part of their academic studies. About 15 teams participate, each composed of 10 students (from different universities) and from different fields (business, engineering, design, etc.) with three of them appointed as business manager, design manager and team leader. This program has given birth to many companies created or co-created by students.

In 2014, a report from the MIT Skoltech Initiative about university-based entrepreneurial ecosystems selected Aalto University as one of the emerging world leaders in this domain (the three other selected cases were Imperial College London, Tomsk State University of Control System and Radioelectronics in western Siberia, and the University of Auckland). Its evaluation of Aalto University is consistent with the results of our own February 2017 analysis: "The student-led entrepreneurship movement emerged in late 2008. It was an anti-establishment movement borne of frustration with the lack of regional and university support for entrepreneurship and a passionate drive to create an alternative, vibrant start-up environment. Started by a small core of students, innovative events and activities quickly engaged the local startup community and attracted a widening pool of students ... Central to this operation is the Aalto Entrepreneurship Society (AaltoES), a not-for-profit student-run society with over 5000 members, drawn largely from the undergraduate and post-graduate population at Aalto and other Helsinki-based universities" (p. 26).

About the role of the university's senior management, this report noted the following: "Despite the university having no explicit

Entrepreneurship and Innovation (EandI) policy, its senior managers have played an important role in creating the conditions for the organic growth of entrepreneurial cultures, activities and communities in and around Aalto" (p. 26). "In particular, their approach has been to 'support but not direct' the student-led entrepreneurial movement, for example by providing public endorsement, financial help and physical space for its activities".

In other words, the university adopted a "hands-offs" approach. The report also indicates that the university "has deliberately downplayed the importance of its IP ownership and startup affiliation, regarding these as secondary to the overarching goal of developing the broader ecosystem" (p. 27).

This 2014 report questioned whether the enthusiasm and volunteerism of the student-led movement would continue after this "highly successful founding phase". One major challenge is the necessary constant renewal of the student leadership. The report suggested the following: "If the Aalto ecosystem is to continue to thrive and progress, the university needs to give priority to embedding an EandI agenda within its core policy, supporting and rewarding its faculty a culture of enterprises and innovation".

In 2017, it was clear that the students themselves continued to provide these entrepreneurial activities. They have managed to develop their activities with the creation of Slush and with the international expansion of their model. But, during this period, Aalto University also became strongly committed to providing an enabling environment for these activities. Notably, university support for this movement involves new programs specifically dedicated to entrepreneurship, including theoretical as well as practical courses. Aalto Venture Program (AVP) is one of the main teaching programs provided by the Aalto University in partnership with the Stanford Technology Ventures Program, centered around entrepreneurship. Any student at the bachelor, master or Ph.D. level from any schools of Aalto University — electrical, engineering, art, design and architecture, chemical engineering, science, business — can take courses from the program, sometimes gaining credits or even taking a minor in entrepreneurship.

In 2017, AVP had 2,100 participants in credit activities and 15,000 participants in non-credit activities. It should be noted that the program is aimed at not only future entrepreneurs but also any students who would like to develop their entrepreneurial skills, regardless of whether they intend to join a start-up or a large company. AVP offers a wide range of opportunities to its students organized around three main themes covering what is needed to properly start a company: "Inspiration" provides students with general guidance about innovation and entrepreneurship as well as extracurricular activities such as Slush volunteer trainings or conferences from eminent and successful start-up entrepreneurs; "Capability" gives the students the concrete tools to get into the venture, with courses about theoretical and practical understanding of the development of a company, storytelling and concrete training relating to real situations; and "Network" represents the access granted by AVP to the whole Aalto ecosystem and network through alumni and mentor's system, global projects, study trips and school exchanges with universities such as Berkeley or Stanford.

Today AaltoES and the university management have achieved a degree of balance between the activities of the student society and the role of guarantor of a favorable environment by the university.

Type 1b: Missionary Professor-Run Entrepreneurship Policy

Even in the absence of university-level strategies to promote student entrepreneurship, some faculty members (sometimes with the tacit or explicit approval of the department) have taken the initiative to support this activity. These faculty members typically hail from business and engineering schools, as well as some science departments. In recent years, the latter have been focusing on faculty and postdoc entrepreneurship. More recently, some of them also include activities targeted toward undergraduates and master's students. Faculty with entrepreneurial experience have played a significant role in supporting other faculty who are new to the spin-off process (Mosey and Wright, 2007). Similarly, such faculty may possess important experientially based skills that can make them more effective as coaches and mentors for student entrepreneurs. Adjunct or visiting professors, entrepreneurs in residence (EIRs) or professors of practice may also be well-placed to perform this role.

The objective of the staff involved in these activities is to develop support mechanisms beyond the traditional classroom offerings. They are bottom-up initiatives even if they can then benefit from the benevolent support of the management of the university. Their focus is on student start-ups. Usually, their resources are limited compared to the integrated approach, which will be presented later. But some of them have, compared to others, important resources. These resources consist generally of one or two missionary teachers, a small space to develop projects and sometimes the financial support of alumni who have created their own companies. Small individual initiatives can develop over time into important entrepreneurship departments. This type tends to show a broad range of activity including entrepreneurship education, competitions, events, conferences, etc. This is the case with some engineering schools, such as MINES ParisTech.

The MINES ParisTech Entrepreneurship Unit

The MINES ParisTech Entrepreneurship Unit (POLLEN for *Pole entrepreneuriat* or entrepreneurship center) is an interesting illustration of this case. This unit was established and led by a professor in this engineering school. This professor acts as a missionary to foster student entrepreneurship in his organization with the approval of the school management but, as an entrepreneur, he has to find his own resources.

Given that MINES ParisTech is a research university, the first step in the development of the center was to capitalize on the faculty's research on academic entrepreneurship and public policy, in order to foster the creation of academic start-up firms. In 2007, a specialist option in Innovation and Entrepreneurship was created (see Chapter 3). This added to the 15 options that already existed at MINES ParisTech: from geosciences to material sciences as well as industrial economics (Mustar, 2009). The professor in charge of this specialization is a member of the Department of Economy, Management and Society. In 2013, POLLEN was created by the Directorate General of the institution.

Its aims are as follows:

- to provide entrepreneurship education for engineering students (innovation and entrepreneurship option), doctoral students and other students of the school (in particular through a series of conferences);

- to support the pre-incubation of start-up projects from engineering students, Ph.D. students, young alumni and staff of the school;
- to help students, doctoral students and staff with a business project benefit from the entrepreneurial ecosystem in which the school participates;
- to better integrate MINES ParisTech into this ecosystem (notably with PSL) and develop exchanges with active universities in these fields abroad;
- to organize conferences, forums and seminars on these issues;
- to organize entrepreneurship training sessions and other activities (expert appraisals, advice or studies) for companies, administrations or associations;
- to set up a network of specialists who can offer services to former graduates who are in the process of creating or acquiring a company;
- to follow all the companies created or initiated by the students, doctoral students, alumni and staff of the school and analyze them in the long term.

The professor in charge of entrepreneurship teaching at MINES ParisTech was appointed head of POLLEN. He remains in charge of the Innovation and Entrepreneurship option. But his position in the organization has changed: he is no longer a member of a department but is under the co-supervision of the Directorate of Research and the Directorate of Education, which indicates the cross-over nature of this entrepreneurship center.

The institution recognizes the importance of entrepreneurship but devotes few resources to it: no budget, no staff, just the time of one professor who was already committed to these activities.

After 4 years of existence, POLLEN defines itself around three main activities:

1. Entrepreneurial training (e.g., the innovation and entrepreneurship specialization).
2. To facilitate start-up creation, thereby to provide care, mentoring and support to MINES ParisTech students and young alumni who are

engaged in start-up creation; to support and provide advice and contacts to early-career researchers or senior scientists from MINES ParisTech seeking to create their own company on the basis of their research results — two or three new ventures are created on average per year.

3. To stimulate and highlight the MINES ParisTech "start-up tribe"; in recent years, the professor has created a series of workshops (the Wednesdays POLLEN), a co-working/pre-incubation space (*La Bulle électrique*) and a start-up competition (*Prix Entrepreneuriat* MINES ParisTech-Criteo); further, he is in charge of a research project focusing on the creation of firms by students, alumni and academics.

The majority of resources come from outside the institution. In 2015, a classroom in poor condition was assigned to POLLEN by MINES ParisTech to be used as a co-working and pre-incubation space. The budget to restore and equip it (nearly €170,000) was raised by the head of POLLEN in equal amounts from the Alumni Association and Paris City Hall, which has launched a program to finance the co-working space between students and entrepreneurs in Paris.

The MINES ParisTech Criteo Entrepreneur Award was created in 2015 with the support of Criteo, a start-up created in April 2005 of which two of the three creators are alumni of MINES ParisTech. Criteo joined the NASDAQ Stock Market, on October 30, 2013. Criteo awards a grant every year to the winner of the prize. Since 2018, the award has included three categories and three awards (€10,000 each): an "Emergence" category for companies created less than 2 years ago by young graduates of MINES ParisTech, a "Development" category for those created more than 2 years ago, and a "Spin-Off of the PSL School of Engineering" category which extends the prize to the ESPCI (*École supérieure de physique et de chimie industrielles de la ville de Paris*) and Chimie ParisTech which, like MINES ParisTech, are very selective and renowned engineering schools.

It benefits from the support and attention of the general management but must find external financial resources to function. The MINES ParisTech Foundation plays a major role for its activities and development.

Type 2: Across the University

Based on our interviews, this model is a basic structure that is encountered in many large and relatively new universities, including those without a long entrepreneurial tradition. Its main objective is to support and develop student entrepreneurship. It is a top-down model, initiated by the management of the university as a means to introduce some support where previously it was absent at the university level but where there may be localized support at the department level. The main focus is to support students and recent graduates who are interested in becoming entrepreneurs.

Typically, a small team is in charge of this activity. The reception space for projects is modest, and there is typically no or little financial support for these projects. These resources tend to be superficial, and there is a danger of trying to do too much with too few resources by trying to provide training programs, mentorship for student entrepreneurs, some access to a space (co-working space), business plan competition, student entrepreneurship clubs, etc. At the same time, there may be conflicts and lack of buy-in by departments who have been developing their own initiatives. As such, while this type may be useful as an initial effort, it may not be sustainable but rather a form of experimentation.

Imperial create lab and enterprise lab

Imperial Create Lab is an example of quadrant 2-type student entrepreneurship initiatives. It illustrates why universities may opt for this type of approach and also the challenges faced.

In 2011, Imperial College London started to explore novel approaches to early-stage investment due to an increasing variety in the entrepreneurial funding landscape. Acknowledging the need to tackle the issue of substantial variation in start-up quality within a given university, a structured pre-acceleration program was developed, called Imperial Create Lab. It was launched as a pilot, under the original name of IC Startup — a student business plan competition — with the clear objective to "skill-up" academic entrepreneurs (no funding attached). Originally established as a development program focused on accelerating a new generation of digital and social-driven start-ups from both faculty and students, it evolved into a pre-accelerator solely focused on deep science, i.e., complex technologies with examples ranging from automated genome editing to an innovative catheter-based system for hemodialysis. This pre-accelerator provided benefits in the form of very hands-on mentoring, access to resources and ongoing monitoring and was characterized by three main elements:

1. An entrepreneurial community platform that generates an ecosystem ensuring diversity by reconfigurable shared spaces and formation of strong ties with particular initiatives that dissolve geospatial boundaries. For instance, it has a formal relationship with Imperial's Advanced Hackspace which is designed for interdisciplinary and interdepartmental collaboration and experimentation by providing social tools, laboratory equipment and advanced manufacturing capabilities.

2. One of the core issues remains at the top of the funnel: pre-idea and pre-team. Tools are developed to form multidisciplinary teams that actually implement solutions to real problems and well-defined challenges through hackathons, event-based activities and partnerships with MNCs such as Google (e.g., solve for X and its breakthrough innovation award).

3. An action-oriented program (duration of 6 weeks) — called Venture Capitalist Challenge (VCC) — tailored to science, technology, engineering and mathematics students, academics and alumni triggering a revolution in (science) venturing by offering a sandbox environment to test high-technology ideas and tackling issues such as a lack of time, space and resources for R&D and prototyping. The VCC was structured in three "selection" phases for selected participants, each phase representing a milestone. In other words, participants need to have theoretical proof and some validated idea of the commercial side before phase 2 and need a working prototype and a deal of some kind to showcase in order to make the transition into phase 3. The final group has the opportunity to pitch before a curated panel of qualified investors during a celebration event. During the program, educational workshops (with topics like funding, customer development, legal, etc.) co-locate in a shared open office space with the aim to encourage collaboration and peer-to-peer learning and sector expert sessions are offered.

The design of Imperial Create Lab was built on an assessment of the regional entrepreneurial environment and the particular needs and expectations of the Imperial community. For instance, together with the direct support of the Business School, it is orienting and even fast-tracking (aspiring) entrepreneurs toward particular types of accelerators that may best meet their needs. Furthermore, it is directly targeting key issues such as IP rights and the visibility of alumni role-model entrepreneurs.

Over 3 years, Imperial Create Lab played a key role in the fundraising of teams (~70 percent) totaling £10 million in proof-of-concept funding, acceleration of 80+ ventures and three acquisitions by the likes of Google and Facebook.

An exemplar of a venture passing through the Imperial Create Lab program is BLOCKS, a wearable technology start-up founded by a team of students from Imperial College London. It is a modular smartwatch designed to give users freedom of choice over the functionality of their smartwatch. Inspired by the idea of modular functionality used by Google Project Ara — a modular smartphone system — the pair took their idea to the Venture Catalyst Challenge, a yearly pre-accelerator program run by the Imperial Create Lab. The BLOCKS team was selected to take part in the 6-week program, during which they shaped their idea into a full-fledged business. BLOCKS took part in the final of VCC 2014, and though they did not win, they subsequently went on to take the Audience Choice award at Intel Wearables 2014 and have gone from strength to strength since. In late 2015, they successfully raised $1 million in a Kickstarter campaign for their first product. It hit the initial funding goal of $250,000 within 1 hour of launch of the campaign.

Despite these positive aspects, several shortcomings of Create Lab called into question its sustainability. It was created under the auspices of a mandate by the College to Imperial Innovations, the College's technology commercialization arm whose remit was primarily the commercialization of faculty innovations as this IP was college-owned. In contrast, the College did not own IP from innovations by graduate and undergraduate students. Create Lab was run by two employees of Imperial Innovations who were effectively "outsiders" who struggled to gain legitimacy from academic departments and the college. The approach was perceived as somewhat ad hoc and below the radar at the college level. Further, there was some concern about whether Imperial Innovations should be involved in these activities as they were commercially oriented. Financing was limited to government funding which placed some constraints on its use for commercialization rather than educational purposes. There were some concerns about whether the initiative was sufficiently impactful at the college level, despite evidence of the number of applications and ventures supported, and at Imperial Innovations, where involvement in student entrepreneurship was questioned. There were also organizational conflicts relating to provision by a commercialization arm rather than integration into the college entrepreneurial and innovation activities.

Because of the shortcomings of Create Lab, Imperial College subsequently decided to create Enterprise Lab in order to centralize all student entrepreneurship activities and put proper management/oversight/key performance indicators (KPIs) in place. This involved a £1.2 million dedicated facility; six full-time staff members dedicated to a variety of flagship programs, such as VCC, Althea (for female entrepreneurs), Idea Surgeries providing general feedback on early stage ideas and "How to" workshops and events; and building an Imperial enterprise network internally and externally (mobilizing alumni). Part of this network is the Experts in Residence (XiR) program, designed to ensure that Imperial-associated individuals including alumni up to 5 years post-graduation, engaged in entrepreneurial projects, and their Imperial start-ups have access to experts who can help them conceive, form, and develop their business. Current experts cover the areas of Legal, IP, Graphic Design, Marketing, Recruitment, PR, and Investment. These individuals act as advisors made available to students via scheduled or drop-in sessions of up to 30 minutes at a time. In addition to this is the Imperial Venture Mentoring Service (Imperial VMS), an advisory service for student or staff with a clear entrepreneurial intent wishing to realize their entrepreneurial ideas through a start-up. Advice is provided free by a pool of mentors who are experienced entrepreneurs. Mentors, who are selected on the basis of them having suitable experience and being suitable to perform such a role, commit on average 1 day per calendar month, meeting the ventures they are mentoring in 2-hour sessions. Efforts are being made to ensure consistency and efficiency by consolidating support so that, for example, business coaching is combined into the XiR program.

These developments bring the student enterprise in line with the central enterprise division within the College. Although organizationally it falls under the provost for research, the key objective is for "enterprise" to become part of the student experience/education offering. An important debate concerned whether a clear line should be drawn between the student enterprise that maximized the student experience and a focus on entrepreneurial mind-set/skills for industry or start-ups, rather than a profit-making objective to launch start-ups which was the logic that Create Lab applied, although some ventures do become successful in raising funds. There remains some debate about the appropriateness of the current emphasis and hence of KPIs in the context of the limited government funding for this initiative and the need for integration with a major college development of a new campus rather than effectively being outsourced. The strategy and nature of provision, therefore, continue to evolve, and we return to this case in Chapter 3.

Type 3: Integrated Type

The feature of the integrated type is a powerful center for entrepreneurship (across all schools and all disciplines). The objective in this model is to provide an internally integrated set of support mechanisms aimed at students within and across the university. The focus is both to educate students in entrepreneurship and to offer them all the necessary resources to realize their project through to start-up. The model is characterized by a high level of resources: a strong and structured team of professionals, different kinds of space (co-working, fablab, incubator, accelerator, etc.) and significant funds from rich alumni, the university and corporates. The MIT Entrepreneurial Center presents a good example of this model.

Martin trust center for MIT entrepreneurship

The MIT Entrepreneurship Center was founded in 1990 (renamed the Martin Trust Center for MIT Entrepreneurship in 2011). It is one of the largest research and teaching centers of the MIT Sloan School of Management. It provides the expertise, support, and connections MIT students need to become effective entrepreneurs and aims to "serve all MIT students, across all schools, across all disciplines". The center was established in 1990 by professors Florence Sender (a serial entrepreneur who taught at the Sloan School of Management) and Edward Roberts (David Sarnoff Professor of Management of Technology). The center very soon was led by the duo: Ed Roberts as chairman and Ken Morse who is a successful serial entrepreneur, playing the role of a managing director. The duo developed a long-term plan for the Entrepreneurship Center, with the objective, among others, to recruit 10 tenure-track faculty and 10 entrepreneur practitioners. In the early 2000s, these objectives were achieved:

> "In 2003, the Entrepreneurial Center's staff and faculty now number 29, with senior lecturers plucked right out of the business world" (*Source*: *ROI*, April 2003, Vol. 5(1), pp. 3–7: The MIT Entrepreneurship Center).

In a special issue of *ROI*[3] in April 2003, Ken Morse gave some clarification about the functioning and objectives of the center:

> "the MIT Entrepreneurship Center is for the whole MIT community — it's not just a Sloan thing. That's one of the reasons we're in E40". Morse said. "To my knowledge, it's the only MIT building that has both engineering and Sloan programs housed together. That's symbolic of what we're trying to do with high-tech entrepreneurship here: It's the interface between management science, technology, and engineering…. A vast number of students come to MIT not imagining themselves as entrepreneurs, but they leave here knowing that it's possible…". The center, continues Morse, "puts students in close proximity to real-world practitioners and faculty who can instruct them in sales and marketing, teamwork, networking tactics, research, building credibility, managing intellectual property and venture financing" (*Source*: *ROI*, April 2003, Vol. 5(1), pp. 3–7: The MIT Entrepreneurship Center).

In 2011, the MIT Entrepreneurship Center received a gift from the Trust Family Foundation. It "donated $10 million in support of

---

[3]*MIT Sloan ROI*, your Return on Investment in MIT Sloan, is a news publication for the alumni, faculty, students, staff, corporate partners and friends of the MIT Sloan School of Management.

construction costs and ongoing programming at the center. These programs include the following: expanded classes on entrepreneurship to reach broader MIT audiences, platforms for student experimentation, opportunities and networking exposure, as well as projects designed to drive entrepreneurship in greater Cambridge and around the world". From that date, the MIT Entrepreneurship Center was renamed the Martin Trust Center for MIT Entrepreneurship.

In his public presentations of the Martin Trust Center for MIT Entrepreneurship,[4] Bill Aulet, who is the current managing director of the center,[5] emphasizes the following role for the center: "Coordination and integration of the decentralized innovation and entrepreneurial ecosystem of MIT".

The website of the center presents its current mission as follows: "The Martin Trust Center for MIT Entrepreneurship provides the expertise, support, and connections MIT students need to become effective entrepreneurs. We serve all MIT students, across all schools, across all disciplines".

This model is integrated because the center supports the five MIT schools: the School of Science, the School of Engineering, the School of Architecture and Planning, the School of Humanities, Arts, and Social Sciences and the Sloan School of Management. We refer to it as integrated because it bridges academic disciplines and because its programs try to foster entrepreneurship across the whole university, including undergraduate and graduate students, staff and faculty.

Based on the 2016 activity report and the center website, we present some of the activities of the Martin Trust Center for MIT Entrepreneurship as follows: courses, people and competencies to support and guide those who have a start-up project, The 100K Competition. The MIT delta V is an educational accelerator.

---

[4]"What is the Martin Trust Center for MIT Entrepreneurship? ... and why is it so awesome?" https://fr.slideshare.net/billaulet/trust-center-overview-presentation-august-2014-v4-38494185.

[5]Bill Aulet has served as the Managing Director of the MIT Entrepreneurship Center since 2009.

Courses

The website of the center offers no less than 63 courses for the academic year 2016–2017. They can be categorized based on their specialization as follows:

- a total of 13 courses focusing on "Foundation Subjects" such as New Enterprises, Business model innovation, Entrepreneurship in Engineering, etc.;
- around 12 courses relating to "Industry Focus" like Energy ventures, Media ventures, Healthcare ventures, etc.;
- a total of 12 courses focusing on "Skill Sets" such as Entrepreneurship and innovation: Legal tools and frameworks, Digital product management, Entrepreneurial finance and venture capital, etc.;
- around 18 courses focusing on "Additional Electives" like Startup sales, D-Lab: Design for scale, Entrepreneurship without borders, etc.

It is noted that these classes "combine theory and practice to give students the opportunity to apply the skills they have learned within the curriculum". The model proposed is a "dual-track faculty" model which brings together academics and practitioners "so that students benefit from a broad range of perspectives and experiences in the classroom".

People as student resources

The Martin Trust Center Staff is composed of 18 people including a leadership team (a chair and two managing directors) and four EIRs. A total of 26 affiliated faculty are linked to the center: 10 professors from the School of Engineering, 13 professors from the School of Management and three professors from the School of Architecture and Planning and approximately 20 lecturers.

The Trust Center provides advice about entrepreneurship and startups on a wide array of topics. The student can ask a question or request a meeting with the EIRs. For specific fields of competence, the EIR can organize a meeting between the student and a professional advisor. More than 30 professional advisors, industry experts and entrepreneurs from a wide range of countries offer their time and expertise to MIT students.

Lastly, the BU Law Clinic is a free clinic offering legal guidance to student innovators and entrepreneurs. It "assists students with a broad range of legal matters related to entrepreneurship and cyber law, from basic issues associated with the founding of start-up companies to novel questions about the application of laws and regulations to students' innovation-related activities".

The 100K competition and other prizes

In 1989, the MIT $10K Entrepreneurship Competition was launched in a joint effort by the Entrepreneurs Club (E-Club) and the Sloan New Ventures Association. By 2006, the event had become the MIT $100K and was established as "the nation's leading business plan competition". It brings together students and researchers from across MIT to launch their ideas and technology into companies. Other competitions exist, such as the Clean Energy Prize, the Creative Art Competition, the IDEAS Global Challenge, Inclusive Innovation Competition and the Lemelson-MIT Student Prize.

The Entrepreneurs Club was founded in 1988 for MIT, Harvard and Wellesley students, faculty, staff, alumni and select professionals. "Members represent a range of experiences and backgrounds including business, engineering, arts and sciences. Many MIT and Harvard start-ups have recruited club members as $100K team members, co-founding partners and equity-sharing employees. The club focuses on helping to develop all aspects of science, engineering and technology business creation". There are several other clubs, usually led by student groups, including the following: The MIT–China Innovation and Entrepreneurship Forum, Sloan Entrepreneurship and Innovation Club, Sloan Entrepreneurs for International Development, Venture Capital and Private Equity Club and Women Business Leaders.

The center allows students from across MIT to have access to a large number of resources.

MIT delta V is MIT's student venture accelerator for MIT student entrepreneurs that prepares them to launch and to accelerate sustainable ventures in the real world. Teams have a dedicated desk space in the co-working space of the center and work on their ventures full-time on their project during the summer. They benefit from coaching sessions and meetings with external mentors. Current MIT students can receive a $2,000 per month fellowship while their project receives funding up to $20,000 during the duration of MIT delta V.

Many other activities are organized or co-organized with the center: $t = 0$ (the time is now) which is a full-day MIT's campuswise celebration of entrepreneurship and innovation, the Entrepreneurship Internship which is a 10-week paid summer internship program connecting MIT undergraduate students with start-ups founded by MIT delta V alumni and ProtoWorks which is the Martin Trust Center's makerspace for students to explore, prototype and experiment their entrepreneurial ideas, etc.

Conclusions

In this chapter, we have developed a typology of different approaches to student entrepreneurship by universities. Although the framework presents "ideal" types, the different approaches are not necessarily mutually exclusive. Universities may evolve between types over time, such as at Imperial College, as they develop more experience in promoting student entrepreneurship and make a stronger strategic/resource commitment to this activity. As we know from faculty-based academic entrepreneurship, it takes time to develop the necessary resources and capabilities to enable their support for entrepreneurship to function (Clarysse *et al.*, 2005). Without such resources and capabilities, universities may engage in extensive activities, but these are unlikely to be effective in generating successful faculty-based spin-off ventures. Our analysis suggests that bottom-up initiatives such as Types 1a and 1b, run by students and missionary professors, respectively, and top-down initiatives such as Type 2 may be effective means of kick-starting student entrepreneurship support but may lack the resources and capabilities to help sustain the student entrepreneurship movement so that ventures can be developed beyond start-up.

In Chapter 6, the concluding chapter, we return to discuss issues regarding implementation of these different types. In particular, we consider issues relating to the successful evolution across types as well as issues relating to potential conflicts between different parts of universities and the university center regarding activities to support student entrepreneurship. Also, we comment on the generalizability of this typology to the broader range of universities including those that are less research active as well as those in emerging economies with severely limited resources.

References

Boh, W. F., De-Haan, U., and Strom, R. 2016. University technology transfer through entrepreneurship: Faculty and students in spinoffs. *The Journal of Technology Transfer*, 41(4), 661–669.

Clarysse, B., Wright, M., Lockett, A., van de Velde, E., and Vohora, A. 2005. Spinning off new ventures: A typology of facilitating services. *Journal of Business Venturing*, 20(2), 183–216.

Mosey, S., and Wright, M. 2007. From human capital to social capital: A longitudinal study of technology-based academic entrepreneurs. *Entrepreneurship Theory and Practice*, 31, 909–936.

Mustar P. 2009. Technology management education: Innovation and entrepreneurship at Mines ParisTech, a leading French engineering school. *Academy of Management Learning and Education*, 8(3), 418–425.

Sala, I., and Sobrero, M. 2018. Games of policy and practice: Multi-level dynamics and the role of Universities in knowledge transfer processes. University of Bologna. Working Paper.

Chapter 3

Entrepreneurship Education

Introduction

In this chapter, we elaborate the element given in Figure 1 of Chapter 1 relating to the role of entrepreneurship education in the student entrepreneurship ecosystem. As such, we draw on existing literature and case illustrations to add richness to the schematic models of student entrepreneurship analyzed in Chapter 2 in terms of both educational content and impact on venture creation. Before we consider recent herculean efforts to promote entrepreneurship education on college campuses, it is important to note that some of the famous successful student entrepreneurs have been college dropouts. Perhaps the two most famous founders/entrepreneurs of the modern computer era fall into the dropout category. As we saw in Chapter 1, Mark Zuckerberg, the founder of Facebook, and Bill Gates, the founder of Microsoft, both dropped out of Harvard to develop their firms. Zuckerberg's path to business success through the venture he developed in his dormitory room on campus, and the role that the university played in that process, was immortalized in the film, The Social Network.

Other successful "college dropout" entrepreneurs who developed their business ideas while attending college include Michael Dell (Dell Computers), Steve Jobs (Apple), Larry Ellison (Oracle), Travis Kalanick (Uber), Evan Williams (Twitter), Jan Koum (Whatsapp), Wayne Huizenga (Blockbuster), Barry Diller (Fox) and Giorgio Armani (Armani). At the graduate level, it is well known that Sergey Brin and Larry Page dropped out of the Ph.D. program in computer science at Stanford to launch Google. At the other end of the spectrum, Richard Branson (Virgin Group) and Simon Cowell (Syco Entertainment) did not even finish high school.[1]

Clearly, entrepreneurship education is not a must to excel. At the same time, we know very little about university or high school dropouts who failed to become successful entrepreneurs. How many were left by the wayside when they could have made it with support? Also, college graduates who go on to become successful entrepreneurs tend to receive less media coverage of their path.

As noted in previous chapters, we believe that entrepreneurship education can enhance both the probability of engaging in entrepreneurship and the success of this activity. A key goal of this chapter is to highlight some highly innovative entrepreneurial initiatives, which attempt to develop students' entrepreneurial competencies, spirit and start-up formation (and thus, constitute "experiential learning", in an entrepreneurial context).

The first section of this chapter reviews evidence on the relationship between entrepreneurship and education. The second section reviews the extent of entrepreneurship education. We then consider research that has assessed the effects of entrepreneurship education to date. Not all entrepreneurship education is the same, so the third section of the chapter delves more deeply into the nature and content of entrepreneurship education and how it has evolved in the last decade and why there has been a shift toward experiential learning. This is followed by a brief discussion of how research on the content, methods and pedagogy of entrepreneurial education has changed our conception of the entrepreneurial process. The final section describes three types of entrepreneurial education initiatives using examples from the U.S., U.K. and France that demonstrate the range of external and internal provision in one dimension, and broad-based general entrepreneurial expertise versus narrow-based start-up experience in another dimension. We conclude with a discussion of the growing importance of experiential learning in entrepreneurship education and the

---

[1] http://www.mytopbusinessideas.com/school-drop-out-billionaires-successful-entrepreneurs/ (accessed July 25, 2016).

implications of this trend for student entrepreneurship for research and practice.

Education and Entrepreneurship

In 2008, two distinguished scholars, Richard Freeman of Harvard University and Ben Rissing of Cornell University, partnered with a leading practitioner on technological entrepreneurship, Vivek Wadhwa of Duke University, to publish an important study attesting to the value of university training for entrepreneurship.[2] The authors surveyed 652 U.S.-born tech founders from 502 engineering and technology companies that were established from 1995 to 2005. These firms had more than $1 million in sales and 20 or more employees. Founders are defined as individuals holding the position of Chief Executive Officer or Chief Technology Officer at the time the firm was established. Around 92 percent of these founders had bachelor's degrees and almost half (47 percent) had advanced degrees (30 percent master's degrees, 10 percent had completed their Ph.D. and 8 percent had an MD or JD).

Not surprisingly, given that the sample is comprised of engineering and technology companies, nearly half of all these degrees were in science, technology, engineering and mathematics (STEM) fields. One-third of the founders had degrees in business administration, accounting and finance, while 6 percent had healthcare-related degrees. The bottom line is that almost all successful tech entrepreneurs were university graduates, indicating they were highly educated with a wide variety of educational backgrounds.

Thus, we are far from the image of the entrepreneur as a college dropout. The general reality is much different. This is especially true, given that Freeman, Rissing and Wadhwa's findings show that the average and median age of key tech founders was 39 when they started their companies (Wadhwa *et al.*, 2008). A total of 95 percent of U.S.-born tech founders were over 25 years of age.

---

[2]This study follows another report from the Kauffman Foundation on "America's New Immigrant Entrepreneurs" (2007).

Consistent with evidence presented earlier regarding the importance of gaining work experience after graduation prior to starting a venture, they find that most founders of technology and engineering companies worked for 16 years before they launched a start-up. This intermediate period between start-up creation and degree completion was shortest for computer science and information technology graduates (14.3-year average) and longest for engineering degree holders (17.6-year average) and for applied sciences graduates (20-year average).

The Freeman *et al.* study also debunks a popular myth that most founders of tech firms graduate from the nation's top research universities. On the contrary, the 628 U.S.-born tech founders providing information on their terminal (highest) degree were educated at 287 unique universities reaching well beyond the likes of MIT, Stanford, Berkeley, Harvard, Columbia, Cornell and Carnegie Mellon. It is true, however, that degrees from top-ranked universities are over-represented in the ranks of these tech founders. For example, while Ivy League universities account for only 1.6 percent of all U.S. degrees awarded, these schools account for 8 percent of the terminal degrees of these founders. Harvard, which was one of the first universities to offer an entrepreneurship course in 1947, dominates the group of Ivy League Schools with 3 percent of the founders.

Ivy League schools are not only disproportionately represented, in terms of degrees, but also appear to generate the most successful companies. In 2005, the average sales and total employment of all start-ups in the sample was around $5.7 million and had 42 workers. Start-ups established by tech founders with terminal Ivy League degrees had higher average sales and employment of around $6.7 million and 55 workers, respectively. A team of researchers from the University of Nottingham: Daniel Ratzinger, Kevin Amess, Andrew Greenman and Simon Mosey, examined approximately 5,000 Internet start-ups and found that founders with a university background were more likely to secure investment funding and successfully exiting businesses (Ratzinger *et al.*, 2015).

Studies such as these demonstrate that university training was an asset to alumni entrepreneurs, many of whom may not start their venture until several years after graduation, but that does not necessarily speak on the effectiveness of entrepreneurship education. This topic is where we turn to next.

Teaching Entrepreneurship

There is a long-standing debate about whether entrepreneurship can be taught. In his classic 1986 book, *Innovation and entrepreneurship*, Peter Drucker neatly debunked the view that entrepreneurship can't be taught, as follows:

Most of what you hear about entrepreneurship is all wrong. It's not magic; it's not mysterious; and it has nothing to do with genes. It's a discipline and, like any discipline, it can be learned.

It is now more than 70 years since Myles Mace taught the first entrepreneurship course in the United States which was held at Harvard Business School in February 1947, attracting 188 out of 600 second-year MBA students.

University education in this field has grown exceptionally fast. As Morris and Liguori (2016) note: "The emergence of entrepreneurship within universities over the past 30 years has been breathtaking. It is not unusual at major institutions to find formal degree programs at the undergraduate and graduate levels (i.e., entrepreneurship majors, minors, concentrations, certificates, master's degrees and Ph.D. programs), a curriculum with 20 or more courses, a portfolio of co-curricular programming, and a broad mix of community engagement initiatives."

Since the 1980s, there has been a proliferation of entrepreneurship courses on college campuses. In 1985, according to the Kauffman Foundation, there were only 250 courses in entrepreneurship in the U.S. As of 2013, more than 400,000 students were taking such courses at 2,600 U.S. colleges and universities. The same explosive growth in entrepreneurship courses at the university level has occurred in all OECD nations. We expect this global growth to continue.

For many authors, this growth in entrepreneurship teaching oftentimes by missionary professors has been too rapid and trainers have not had time to develop their courses in a sufficiently reflexive way. Morris and Liguori (2016) comment that while this growth has given a large number of young entrepreneurs a set of tools and concepts that can enhance their likelihood of success, and encouraged the development of ecosystems to support an entrepreneurial community, it has occurred so rapidly that it has outpaced understanding of what should be taught, how it should be taught and how outcomes should be assessed.

Many elite universities, which once thumbed their noses at entrepreneurship, are offering courses and programs in entrepreneurship and are even allocating space to entrepreneurial activity. Natasha Singer of the *New York Times* reports that Princeton University, which does not even have a business school, offers a variety of entrepreneurship courses (Singer, 2015). In the same article, it was reported that Princeton has dramatically increased space allocated to student incubator and accelerator programs to match rival institutions, such as Cornell, the University of Pennsylvania, UC Berkeley, Harvard, Stanford, Yale, NYU and Columbia.

Growth in the number of entrepreneurship courses has largely been driven by increases in student demand, in part because many traditional career paths (e.g., working for a large multinational corporation) are no longer easily available. As such, entrepreneurship education has become an important topic for both public officials and universities because of the perceived impact of entrepreneurship on economic growth and employment. Students are also taking advantage of special programs and financing, the aforementioned incubation space and new technologies that have made it easier to launch a start-up firm.

The rise of the Internet has greatly reduced start-up costs and entry barriers. The concomitant widespread digitalization of the economy has also had a strong impact on student start-ups. On the one hand, this impact arises through the expansion of the creation or discovery of business opportunities (e.g., the development of the sharing economy) and on the other by making opportunity pursuit/firm creation easier (e.g., free turn-key website, free software bricks, social networks for marketing, secure online payment facility and cloud storage services for data (see Hayter, 2016)). The Cloud has also made it much more feasible for students to create start-ups. For example, a modern app company does not require servers and time to develop distribution channels, thanks to various innovations in the online space.

Given the popularity of entrepreneurship courses, it is not surprising that there has been a concomitant substantial growth in the number of entrepreneurship professors in business schools. As of 2018, the Entrepreneurship division of the Academy of Management has approximately 3,285 members. It is one of the fastest growing and largest divisions of the Academy, which has approximately 20,000 members in total. Note that this does not include professors in other fields of business administration who might not belong to the Academy, but who teach courses and conduct research on entrepreneurship (e.g., finance and marketing professors). The rapid growth of entrepreneurship journals is also a testament to the importance of this topic. As an important indicator of quality, there have also been more papers on entrepreneurship in established top journals in recent years.

International Benchmark of Entrepreneurship Education

We conducted an in-depth analysis of the entrepreneurship education policy, programs and content in the following institutions: MIT, Stanford, Cambridge, Imperial College, King's College, Aalto University, Technical University Munich, ETH Zurich, Michigan State University, Centrale-Supelec, ESCP Europe, ESSEC, HEC and the University at Albany, SUNY. This analysis yielded the following insights:

- Entrepreneurship training can be found everywhere: in universities, business schools, engineering schools, etc.
- These courses are aimed at a wide diversity of audience (undergraduate and graduate level) and of forms (minors, majors, masters, initiation, deepening, etc.).
- Instructors are generally professors who research and publish in these fields. This does not prevent them from involving professionals such as former students who have become entrepreneurs, sector experts, venture capitalists, etc., who give seminars and act as mentors.
- These institutions generally have a very broad definition of entrepreneurship that is not limited to business creation.
- The teaching methods are diverse: courses, case studies, professional seminars, group work, progressing from idea to project, business model development, etc.
- In many cases, there has been a shift from traditional teaching methods to experiential learning where students develop a business project.
- In some universities, as we saw in Chapter 2, this movement requires a very strong mobilization of student clubs and also of the broader ecosystem (regions, alumni, property-based institutions, such as incubators, accelerators and science/technology parks).

Entrepreneurial Education Effectiveness

The substantial rise in the number of entrepreneurship courses and programs has generated considerable debate about the effectiveness of formal entrepreneurship instruction. The conventional wisdom is that entrepreneurship programs help develop students' entrepreneurial intentions and to get the ball rolling on their ventures. There is something of a consensus among scholars that entrepreneurship education programs increase the probability that students actually engage in entrepreneurship or increase their intentions to do so. This rapid and global growth has led to much work on the effects of entrepreneurship education. We present below the results of three recent literature reviews on this topic. Each provides insights that are particularly relevant for practitioners, university administrators and policymakers seeking to develop student entrepreneurship.

The link between impact and types of entrepreneurship education and training

Bruce Martin, Jeffrey McNally and Michael Kay conducted a systematic analysis of 42 independent samples (Martin *et al.*, 2013). The departure point of their study is the fact that qualitative reviews have been equivocal about the impact of entrepreneurship education and training (henceforth, EET). They report that most studies show positive relationships, but a number of important studies have shown negative results for EET on entrepreneurship-related human capital assets and entrepreneurship outcomes. In their quantitative review of the literature, they distinguish two kinds of impact: (a) impact on entrepreneurship-related human capital assets (e.g., entrepreneurial knowledge and skills, positive perceptions of entrepreneurship, intentions to start a business) and (b) impact on entrepreneurship outcomes (e.g., nascent behaviors, such as writing a business plan and seeking funding, start-up, entrepreneurship performance which comprise financial success, duration of running a business and personal income from owned business).

Based on their review of the literature, they divide the type of EET into two categories: (a) a relatively short training course that focuses on core entrepreneurship knowledge and skills related to starting a particular business in a particular jurisdiction, and (b) full academic courses that provide a broad theoretical and conceptual understanding of topics, such as how opportunities are identified, decision-making in highly ambiguous contexts, and causation and effectuation. They posit that the differences between these two main types of EET will influence outcomes differentially.

Their results show that both types of EET — training-focused educational interventions and academic-focused educational interventions — have the same effects on human capital assets related to entrepreneurship. But, what is more surprising is that, they also show that the relationship between EET and entrepreneurship outcomes is stronger for academic-focused EET interventions than for training-focused EET interventions.

This last point is important because it suggests that training-focused programs may benefit from the introduction of more conceptual material, which may help students achieve financial success and maintain a business over an extended period of time. Thus, an important lesson for all those who want to develop student entrepreneurship by developing entrepreneurship education is that introducing recent academic findings or questions into courses can be an advantage for the success of their venture. More broadly, this raises the crucial question of the relationship between entrepreneurship research results and entrepreneurship education.

Environment and institutional context matter

While most studies show a positive effect between entrepreneurship education (EE) and entrepreneurship intention, some find a negative effect. These mixed results suggest that the environment has a role to play in the effects of EE. In their study, Walter and Block (2016) explore how the relationship between EE and entrepreneurship activity is conditioned by a country's institutions. Their study is of particular interest since most research on the effects of EE has been limited to the individual level and has neglected country-level contextual influences. Their multilevel analyses of data from more than 11,000 individuals in 32 countries show that entrepreneurship education is more effective in relation to stimulating more entrepreneurial activity, in countries low in entrepreneur-friendly regulations, financial capital availability, control of corruption and public image of entrepreneurs. In institutional environments that are entrepreneur friendly, individuals can acquire entrepreneurial motivation and skills from sources other than EE but not in entrepreneur-hostile institutional environments.

These findings provide a further lesson for those seeking to develop student entrepreneurship because they show that entrepreneurship training courses need to be adapted to the context in which they take place. As there is no "one size fits all", a type of training relevant in one context is likely not relevant in another.

Relationship between pedagogical methods and specific outcomes

Nabi *et al.* (2017) systematically review empirical evidence on the impact of entrepreneurship education (henceforth, EE) in higher education on a range of entrepreneurial outcomes, analyzing 159 published articles from 2004 to 2016. The authors note that research on the effect of EE is mainly concerned with "entrepreneurial intentions" and more rarely the creation of companies and their performance. Such outcomes generally focus on the short term, and more rarely on the long term. Lastly, studies tend to provide scant detail on the actual pedagogies being tested. Bearing in mind these limitations, most of the available research highlight a positive impact (on entrepreneurial attitude and business creation), but some show ambiguous results (e.g., training can have discouraging effects because it gives students a more realistic idea of the difficulties of creation).

Nabi *et al.*'s review calls for "less obvious, yet greatly promising, new or underemphasized directions for future research on the impact of university-based entrepreneurship education. This includes, for example, the use of novel impact indicators related to emotion and mind-set, a focus on the impact indicators related to the intention-to-behavior transition, and exploring the reasons for some contradictory findings in impact studies including person-, context-, and pedagogical model-specific moderators".

Other systematic reviews of the vast empirical evidence by Pittaway and Cope (2007) and Bae *et al.* (2014) confirm these points. Colombo and co-authors report that alumni of technology-based universities are more likely to become entrepreneurs if they complete economics and management courses, with a more specialized course curriculum (Colombo *et al.*, 2015). Rauch and Hulsink compared students taking a master's program in entrepreneurship with students taking a master's program in supply chain management and found that entrepreneurship education was effective in increasing attitudes and entrepreneurial intentions (Rauch and Hulsink, 2015). Souitaris, Zerbinati and Al-Laham found that, in science and engineering, the most notable benefit of entrepreneurship programs is that it enhances students' enthusiasm regarding entrepreneurship (Soutiaris *et al.*, 2007). Wilson, Kickul and Marlino reported that entrepreneurship education can positively affect student beliefs that they can become entrepreneurs (Wilson *et al.*, 2007). DeTienne and Chandler found that entrepreneurial education can also help student entrepreneurs with opportunity recognition skills (DeTienne and Chandler, 2004).

There is more debate, however, regarding the impact of education programs on the level of graduate entrepreneurship and on whether such programs enable graduates to become more effective entrepreneurs. Importantly, impact relating to business start-up and growth may not occur until sometime after graduation (Nabi *et al.*, 2017). Hence, as we showed in Chapter 1, it is necessary to analyze recent alumni, not just current students.

Effectiveness is not necessarily a mere matter of instruction. Another dimension to consider is the extent to which students who already have entrepreneurship experience influence the entrepreneurial behavior of those students seeking to become entrepreneurs, but who do not yet have entrepreneurial experience. The presence of classmates who are experienced entrepreneurs may help inexperienced students identify the most promising opportunities as well as create awareness of the real challenges in starting a venture, which may in turn enable the least skilled and committed students in deciding on whether they would like to become entrepreneurs.

Lerner and Malmendier investigated this issue of peer effects by analyzing 5,897 Harvard MBA students over the period 1997–2004. They found that exposure to a higher share of peers with a pre-MBA entrepreneurial background led to *lower* rates of entrepreneurship, post-MBA, and that this is largely attributable to intra-class learning from experienced entrepreneurs (Lerner and Malmendier, 2013). These findings suggest that the benefit of exposure to entrepreneurship helps both in ruling out ventures that seem likely to fail and in weighing the real challenges of starting a venture. There may be a need for support for would-be entrepreneurs to critically evaluate their business ideas, as well as to downplay the emphasis on business plan competitions.

Assessments of the impact of education on student entrepreneurship should be viewed with some caution. It is possible that the positive effects of education may be overstated, due to the methodological weaknesses of some of the studies (Martin *et al.*, 2013). One problem is "right-censoring" of the data, i.e., studies need to allow sufficient time following a program for entrepreneurs to actually undertake actions to create their ventures, which may take anywhere from 18 months (Rauch and Hulsink, 2015) to 8 years (Kolvereid and Moen, 1997).

Another issue relates to content. Academically oriented courses provide a broad theoretical and conceptual understanding of topics, such as how opportunities are identified and how to make decisions in highly ambiguous contexts. Such tools acquired in these academically oriented courses may transfer better to entrepreneurial outcomes like financial success and business survival than tools acquired in more hands-on focused entrepreneurship courses, which are aimed at instilling core entrepreneurship knowledge and skills related to starting a particular business, since entrepreneurial contexts require decision-making in highly uncertain and dynamic conditions.

Notwithstanding this caveat, what insights can we glean from the available set of studies about the effectiveness of entrepreneurship education?

There are several key stylized facts regarding the empirical evidence. First, there is a difference between an intention to start a venture and actually doing it. Many students may intend to start a venture and indeed go through some of the activities associated with doing so but, in fact, do not actually start one. There is strong evidence that entrepreneurship education has a positive influence on the *intentions* to engage in entrepreneurship. Unfortunately, the evidence is much less clear on the effects of entrepreneurship education on entrepreneurial *actions*. Second, the context in which entrepreneurial education is presented may adversely affect students' intentions and possibly subsequent actions, such as a weak university entrepreneurial culture and students with previous entrepreneurial exposure (Nabi *et al.*, 2017; Wang and Verzat, 2011; Lerner and Malmendier, 2013).

Third, it appears that elements of an entrepreneurial ecosystem are also critical. For example, while individual characteristics and education programs are influential, importance is also attached to the availability of incubator space and interaction with those who have entrepreneurial experience. There is a need to understand the factors that shape such an ecosystem, the role that these factors play and the elements that can impede the development of student start-ups (Van de Ven *et al.*, 1999). Hence, more entrepreneurship education as such may not necessarily be beneficial, but rather how that education relates to other elements of the ecosystem.

The Content of Entrepreneurship Education

Thus far, we have considered the proliferation and general shape of current programs. However, as an academic discipline, entrepreneurship has been taught for well over 40 years. Over time and especially in recent years (as more full-time academics are teaching the subject), its methods and objectives have considerably evolved.

Although in his review, Morris (1998) identified 77 different definitions of entrepreneurship, in essence, there are two approaches to entrepreneurship in academic literature: a narrow one and a broad one. Within the narrow view, entrepreneurship is the creation of organizations or the process by which new organizations come into existence (Gartner, 1988). Entrepreneurship is about individuals creating new enterprises, starting up new businesses. With this view, entrepreneurship is linked to start-up creation, but also to family business and self-employment.

The broad approach to entrepreneurship is to consider the creation of new activities or value in multiple organizational contexts. Entrepreneurship is about how, by whom and with what consequences opportunities to bring future goods and services into existence are discovered, created and exploited (Venkataraman, 1997). Based on this broader perspective, entrepreneurship is not only about start-up creation. It can occur in existing organizations (intrapreneurship, corporate entrepreneurs, organizational renewal, etc.). Also it is viewed as a way of thinking, of acting, and it is the nexus of individual opportunity (Shane and Venkataraman, 2000), and can be linked to initiatives, innovation and opportunity orientation.

These distinctive views about entrepreneurship have important implications for the goals and outcomes of entrepreneurship education. If entrepreneurship is considered only as starting a business — the narrow definition of entrepreneurship — the goal of entrepreneurship education is to deliver the knowledge and know-how to create new firms and to foster the ability of students to start their own company. If entrepreneurship is considered more broadly, the goal of entrepreneurship education is to foster creativity, to develop entrepreneurial spirit and behavior, and to help students with capabilities to identify and exploit opportunities and ideas that create value (economic, social, cultural and financial). The former dominated earlier developments in entrepreneurship education. The latter has come to the forefront more recently, broadening out the scope and applicability of entrepreneurship education as it became clear that there was a demand for students to be equipped with these entrepreneurial skills in an increasingly competitive and dynamic work environment. Even if they did not have the intention to start a venture, such skills were important in a corporate entrepreneurship setting as well as in relation to the eventual possibility that graduates might seek to acquire a business.

For a long time, the two main methods used in entrepreneurship courses were case studies and the formulation of lengthy and somewhat tedious business plans. Increasingly, the effectiveness of these approaches has been called into question as the nature of entrepreneurial ventures has changed, especially with respect to the amounts and nature of start-up funding they require. This has especially been the case with the growth of ICT-type ventures. There is therefore less need to produce the kind of detail required by venture capital firms. Rather, there is a need for students to actively shape the business models for their venture ideas rather than construct abstract business plans as a classroom exercise.

New teaching formats therefore actively engage students in the courses. The techniques for teaching entrepreneurship include starting businesses, serious games and simulations, design-based learning and reflective practice (Neck and Greene, 2011). In recent years, pedagogy has shifted toward a stronger emphasis on experiential learning. That is, the goal of entrepreneurship education is no longer "to learn how to create a start-up", but rather it is "to create a start-up". Over the past decade, learning by doing (e.g., starting a business) has become the focus of coursework. In the majority of the "new" courses, students have to start a business to understand real-world entrepreneurial practices. With these new formats, entrepreneurship education is not limited to acquisition of knowledge but also to experimentation where the students deal with scarce resources, with a value proposal, with future customers, with a business plan, with partners, etc. (Mustar, 2009). The skills and competencies acquired by students through these formats are not only useful for setting up their own business but also to act entrepreneurially in various other contexts.

Schematically, entrepreneurial education has left behind a "classical" model comprising different phases that go linearly from idea to success. In the classical model, a predictable entrepreneurial process is as follows:

1. You have a brilliant idea (you have discovered or created an opportunity).
2. You begin to write a business plan.
3. You engage in fund-raising (e.g., family, friends, angel investors, VCs).
4. You hire a great team.
5. You build the product or service of the new firm.
6. You sell it to the customers.

With this model, two things are crucial: the starting idea and the money. With the "new" model of the entrepreneurial process, execution is more important than conception of the first idea or the first opportunity, given that entrepreneurial teams will inevitably transform this idea several times during the creation process. Also, entrepreneurs typically do not require a substantial amount of money to develop their project because they bootstrap constantly.

This "new" model has been popularized to a wide international audience by many authors on the fringes of the academic world. Steve Blank's book *The Four Steps to the Epiphany* which was designed as a companion to his entrepreneur class at the Haas School of Business (UC Berkeley) was self-published in 2003. In his book, he developed the Customer Development methodology, which launched the Lean Startup movement popularized by Eric Ries. Eric Ries's book *The Lean Startup: How Today's Entrepreneurs Use Continuous Innovation to Create Radically Successful Businesses* was published in 2011. Another book used today by hundreds of thousands of start-ups and entrepreneurs worldwide is Yves Pigneur and Alexander Osterwalder's bestseller *Business Model Generation*. The tools, methods and concepts developed by these authors and others have had and continue to have strong effects on the start-up phenomenon and education. They propose new and dynamic approaches to business modeling.

Types of Experiential Entrepreneurship Initiatives

There is little published research that provides concrete examples of the activities carried out by students during experiential learning. One of the rare studies is by Mason and Arshed (2013) who describe in detail a first-year entrepreneurship course in a Scottish university using this approach. Their case study shows that the experiential learning assignment "was an effective learning experience for the students, complementing and rein-forcing prior classroom learning through application. It facilitated learn-ing about the real world of the entrepreneur, something which would otherwise not have been possible, and had a positive impact on entrepre-neurial intentions".

However, it is clear that experiential learning initiatives do not involve a "one-size fits all" approach, even within the quadrants we identified in Chapter 2. To illustrate the importance of experiential learning, we focus this section on examples of three types of entrepreneurial initiatives that attempt to develop entrepreneurial competencies, as well as an entrepre-neurial spirit in an experiential setting. The first type of initiative involves an externally driven co-curricular program heavily focused on providing a range of mentoring and support activities to enable students with a clear entrepreneurship idea to take it to the start-up phase. Interestingly, this is a learning concept that is being rolled out nationwide across many universities and which also has a policy objective to have multiplier effects beyond individual universities to local communities, through job and other business creation. The second type is an internally driven educational program involving integrated classroom and experiential learning resulting in the development of a business concept. The third type is a more broad-based program to stimulate entrepreneurial skills and thinking in relation to the commercialization of innovations through both start-ups and corporate entrepreneurship.

Externally Driven Co-Curricular Start-Up Program: Blackstone Launchpad

An innovative student entrepreneurship initiative that is driven externally is the Blackstone LaunchPad program in the U.S., which is supported by the Blackstone Charitable Foundation. Blackstone is one of the world's largest private equity firms. The Blackstone LaunchPad program was originally developed at the University of Miami and is now serving more than 500,000 students at 20 colleges and universities nationwide.

This nationwide network now includes centers in the five largest states in the U.S. (by population), namely California, Texas, New York, Florida and Pennsylvania. In 2015 and 2016, Blackstone established new centers in New York and Texas. The universities selected in New York were Cornell, NYU, Syracuse, University at Buffalo, SUNY and University at Albany, SUNY. In 2016, three Texas universities were added: Texas A&M University, The University of Texas at Austin and The University of Texas at Dallas.

Blackstone Launchpad is a co-curricular, experiential education program, which introduces entrepreneurship as a viable career path to all students at a college or university, regardless of their major, level of experience or academic discipline. Many entrepreneurship professors (both adjunct and full-time professors) are involved in these centers, so they bring a strong educational background to the program. Blackstone LaunchPad is designed to support and mentor students, staff and alumni, while, at the same time, developing entrepreneurial skills and an entrepreneurial mind-set. Students have access to a network of venture coaches and an entrepreneurial support system.

A key objective of the program is to develop new ventures. The incubation process proceeds as follows:

1. A student entrepreneur completes a profile, which can be accessed by all members of the LaunchPad Community. They also complete a "Venture Assessment Form", which should outline the details of their expected market, competitors and the future of their product or service.
2. LaunchPad staff assign each prospective student entrepreneur with a "master" entrepreneur, who helps that student develop the business idea. These experienced "master" entrepreneurs can recognize great potential and they guide students through the rigorous process. The prospective student entrepreneur also has access to other members of the LaunchPad staff.
3. The LaunchPad does the networking and provides the student with the proper resources. Each prospective student entrepreneur also can meet with business leaders in their region for additional mentorship. Partnerships are also formed with programs, firms and investors in the area.
4. Blackstone helps bring their product to the market, and this creates new jobs in the area, leaving a lasting, sustainable network.

One of the first Blackstone LaunchPads was established at the University of Central Florida in Orlando. UCF is one of the fastest growing universities in the country. An interesting example of a successful Blackstone student start-up was the one established by Jesse Wolfe when he was a UCF senior. His firm is called O'Dang Hummus. The major thrust of his business plan was to differentiate his firm. He noted that all the major firms in the market were offering the same flavors. His idea was to "disrupt" the industry, by creating a variety of new flavors of hummus, such as Sweet N' Spicy Black Bean and Bombalicious Buffalo Wing. He believed that such flavors would appeal to younger consumers. His start-up was the highest-ranked student-led venture in the inaugural Blackstone LaunchPad Demo Day held in New York City in 2014, winning $15,000 for further business development. O-Dang Hummus has since greatly expanded nationally. Jesse appeared on the famous TV show, Shark Tank, and his firm is now also producing salad dressings.

Internally Driven Educational Program Involving Integrated Classroom and Experiential Learning: The Entrepreneurial Journey Course (Imperial College Business School)

The focus of the Imperial College Business School's enterprise approach is to equip students and academics with an entrepreneurial mind-set (equally relevant to starting-up, scaling-up or steady-state organizations), from undergraduates to post-graduates and researchers. This is done by embedding entry-level and state-of-the-art enterprise education in experiential programs. This targeted approach helps students and academics to limit resources and increase the impact.

Initially called the Innovation, Entrepreneurship and Design (IED) course and now relabeled the Entrepreneurial Journey, this is the school's flagship enterprise education program and involves both the full- and part-time MBA programs. This course aims to enhance the entrepreneurial culture at Imperial College by giving students the opportunity to explore entrepreneurial ideas and develop practical entrepreneurial skills. The focus has been placed on the exploration of opportunities offered by a learning-by-doing approach. It is distinctly designed from traditional entrepreneurial course interventions as it combines different sets of learnings and tools aimed to instill a sharp customer focus, an agile mind-set and a clarity around execution milestones. This has led to a model of how the likelihood of positive outcomes can be improved upon in the formative phases of ideation so that innovative concepts can be transformed into feasible business cases. We have identified the following key differentiators:

- The central role of the research-driven Innovation and Entrepreneurship Department at ICBS and its world-leading research. It informs and develops the program, bringing together innovative thinking and insight with new technologies.
- A purpose-built virtual learning platform that has been developed to provide specific targets and helps clear the barriers to success. This platform is closely linked to a book *The Smart Entrepreneur* (Clarysse and Kiefer, 2012). The platform provides a structured approach in

addressing the key elements in framing an opportunity in its potential market and establishing a venture.

- The goal of the Entrepreneurial Journey is to produce a final business pitch — based on a slide deck — that is presented to a panel of investor practitioners and is considered for a £5,000 prize. For the academic assessment, students have to deliver the following:
  - A practice pitch that takes place a few weeks before the final pitch.
  - An expanded pitch deck that is an expanded version of the final business pitch slide deck designed so that it can be sent to potential investors, and should speak for itself.
  - A detailed financial spreadsheet accompanying the expanded pitch designed so that it can be sent to potential investors, and should speak for itself.
- Workshops to kick start the idea generation and formulation process comprising a Design Thinking workshop, a Lean Innovation workshop (provided by the Enterprise Lab), and a Get Out of the Building workshop to identify customers' pain points and validate assumptions on customer problems.
- Formal coaching where teams meet up 3 times a month for 45 minutes with an entrepreneurship faculty member coach to present their ongoing work and receive feedback for the further steps covering Idea Pitch, Value Proposition and Market Feedback. Events organized by the Business School giving opportunities to connect with other students, share or view ideas and form teams before the core module begins, notably are as follows:
  - Idea Pitch and Team Formation events to share idea or skillsets within the MBA cohort with a view to networking and forming teams.
  - Idea surgeries — these are an opportunity to meet a faculty coach from the module to discuss your business idea at an early stage and get feedback.
  - HootBoard, an online platform provided in partnership with the Enterprise Lab to share or view ideas and connect with other students.
  - Idea Marketplace, a showcase event providing opportunities to meet "non-MBA/external" students and postdocs from other

Imperial departments who may be ideators for new products or services, or may provide a valued skillset relevant to the start-up idea.

o Pitch n Mix events, which are the Enterprise Lab's monthly events that bring together like-minded students to hang out, get feedback and meet new people. Students have the opportunity to participate in the Imperial Business Pitch competition following the module. This is a business plan competition consisting of three rounds. First, by submission of an online pitch video, including details of the market problem and opportunity, an assessment is undertaken by a panel of external judges, i.e., qualified investors. A selection is made of a long list for extra mentoring and support. Second, on the day of the final, the selected teams will pitch in front of a group of experienced venture capital investors and serial entrepreneurs who decide which teams go forward to the final. Third, the finalists will compete in front of the audience and a panel of high-profile Dragons' Den/Shark Tank-style judges. Reflecting the evolution of the initiative, the Imperial Business Pitch competition is now open for all Business School programs and not just for this module specifically.

In sum, the course is primarily an entrepreneurship education with a strong focus on learning-by-doing, which incorporates business generation as a side effect.

The existence of this course is a major reason why many MBA students come to Imperial and it makes a strong contribution to Imperial which has been ranked 2nd for entrepreneurship in 2019 in the QS MBA ranking (*Source*: https://www.topmba.com/mba-rankings/qs-global-mba-specializations-rankings-0/out-now-qs-mba-specialization-rankings-2019). Because of its success, the entrepreneurship model has been transcribed and integrated in the curriculum of programs across the Business School and Imperial College. For instance, based on the main principles, an entrepreneurial boot camp has been designed for international students with the aim to develop their entrepreneurial competencies. This intensive boot camp takes place each summer and is delivered in such a way that the whole content is condensed into 3 weeks instead of 6 months.

An example of a student-based enterprise emerging from the module is Breathe Easy. Breathe Easy was brought to market by a multidisciplinary team of four MBA students, an Electrical Engineering student from Imperial College and an experienced Product Designer and Entrepreneur from the Royal College of Art. It is an innovative product designed to enhance indoor air quality by cleaning the air of toxins or volatile organic chemicals caused by synthetic materials used in the construction of buildings.

Breathe Easy uses toxin absorbing plants, as researched by NASA, with a novel self-sufficient hydroponic extractor fan system, which provides the plants with the conditions required for healthy growth and extends their toxin-absorbing range. Breathe Easy was one of the five projects that entered into the 2010 Innovation, Entrepreneurship and Design Business Plan Competition and was awarded the first prize of £10,000.

Charis Hewitt (full-time MBA 2010), lead entrant, reflected on the learning process as follows: "The course really taught us the process that you should go through if you ever want to set up your own business. It's taken us from the step at which you have a simple idea, shown how you should evaluate that idea, taught us how to do market research, how to evaluate who you should be targeting, how to position your product and how to put together a strategy for how you can sell your product".

Broad-Based Program to Stimulate Entrepreneurial Skills and Thinking: Innovation and Entrepreneurship Specialization (MINES ParisTech)

Higher education in France is characterized by two types of institutions: the universities and specialized schools, which are mainly in engineering and management. After the equivalent of junior high school in the United States, French students with the strongest academic credentials attend preparatory classes for 2 years to prepare for a very difficult and highly competitive selection process for admittance to the most prestigious Grandes Ecoles. MINES ParisTech is one of the most famous and highest ranking French "Grandes Ecoles" in engineering (HEC in Paris is ranked as the best French "Grande Ecole" in management).

Since its founding in 1783, MINES ParisTech has educated engineers capable of solving complex problems in a diverse array of fields. MINES ParisTech is one of the nine full member institutions of Université PSL (Paris Sciences & Lettres), a university formed in 2010. PSL ranks 41st in the 2018 Times Higher Education world rankings, allowing France to return to the top 50 global universities.

Over 3 years, MINES ParisTech provides training for non-specialized, high-level engineers who will hold positions of responsibility in a variety of fields, such as production, RandD, consulting and management. Several CEOs of the 40 leading French companies (CAC 40) are former graduates of MINES ParisTech. At the end of their 3-year program, a small number of students (around 140 graduates per year) receive a Master of Science in Executive Engineering degree from MINES ParisTech.

The course objective is to ensure a sound common scientific culture in the fundamental disciplines (e.g., mathematics, physics, mechanics and economics) and to enable students to extend their knowledge in a field of their choice. Students choose from 15 possible specializations, one of which is Innovation and Entrepreneurship.

The Innovation and Entrepreneurship Specialization (IandE) prepares engineering students to create a start-up, based on an innovation developed at the university. Such activities may lead to the creation of start-up firms or entities within existing companies (i.e., corporate entrepreneurship). This specialization provides students with important skills that aid in the commercialization of innovations, such as identifying, creating, and exploiting opportunities, managing uncertainty, communicating, building and leading an entrepreneurial team, marketing, entrepreneurial finance, innovation management, business models, and managing intellectual property.

Unlike conventional academic programs, the IandE program begins with an immersion project in a vibrant entrepreneurial environment. During their second year at MINES ParisTech, about 10 students who choose the IandE and their professor undertake a 2-week mission abroad. The goal of this travel is to study an entrepreneurial ecosystem. In the past 4 years, the group visited New York City, Shanghaï, Berlin, Helsinki, London and Cambridge. During these 2-week visits, student interviews take place all day and most of the evening. They are quite intense.

The group meets with a variety of players and institutions associated with the entrepreneurial ecosystem, including visits to numerous start-ups, firms on incubators and accelerators, angel investors and venture capitalists. They meet dynamic entrepreneurs who transmit their passion to them. Following this excursion, students write a report.

The third year and final year of the MINES ParisTech program are mainly dedicated for specialization, which occupies three main periods in the academic year: (a) 4 weeks full-time in October, (b) 3 or 4 weeks full-time in January and (c) 3–6 months full-time from the end of March to the end of September. Between these three periods, the students take other courses in fields outside innovation and entrepreneurship. They also devote 1 day a week for the IandE.

The October and January periods

The October and January periods are organized around the creation of an actual start-up, involving groups of two or three students. The students must define a target market, identify future customers or users, analyze the external environment, identify key resources and key activities, meet potential partners, and develop a viable business model.

This exercise is an opportunity for students to directly confront the realities of entrepreneurship and work with key players in this process. During these periods, students occupy a specific space within MINES ParisTech (both co-working space and pre-incubation). The professor in charge of the IandE meets with students daily, in his office, which is adjacent to where the students are based. The professor interacts with these students, answers their questions and provides them with suggestions for ideas and contacts.

The project is conducted in parallel with teaching modules and workshops that take place in this co-working area. Schematically, students spend half of their time on the project. The other half includes two sets of activities that are as follows:

1. Teaching modules with professors (marketing of innovation, start-up strategy, finance and financing, and legal and regulatory issues).

2. Meetings and workshops with entrepreneurs and professionals (venture capital, design, IP, web programming, incubators, social entrepreneurship, start-ups in clean tech, in biotechnology, etc.).

On a typical day, students take an entrepreneurial finance course in the morning, which is taught by a professor from HEC (a leading business school in Europe). Next, they spend 4 or 5 hours on their project. In the evening, an entrepreneur delivers a presentation on his/her start-up or the group will participate in a workshop with, for example, a specialist Internet business model. Such meetings with many diverse people allow students to accumulate knowledge and contacts, but also helps them address issues more directly related to developing their own project.

In addition to the professor in charge of the training and alumni of this specialization, different external mentors meet regularly with students and help them develop or accelerate their project which is, in February, presented to a real investment committee.

The end of March–end of June period

After the investment committee presentation, the students and their professor decide either to continue their "start-up creation project" and turn it into a real venture, or to stop this pedagogical project and spend 3–6 months working in an entrepreneurial situation in a start-up, a VC firm or a non-profit. After this period of internship, they are awarded a diploma.

Up to a dozen students take this specialization each year. At least one-third of the students decide to continue their project and remain in the pre-incubation space at MINES ParisTech. The others complete the end-of-course internship in a start-up or in an organization outside MINES ParisTech.

In 2016, two groups of two students decided on their end-of-course internship and tried to transform their start-up project into a real company: Q4U is a multiplatform start-up for improving the user experience of waiting in queues (in supermarkets, hospitals, etc.). Q4U works by allowing a user to join a queue through their smartphone. As such, the user can move around freely during the waiting time and avoid the disadvantages of a physical queue. Q4U is also an asset for companies and partner organizations since this solution is both a user data analysis tool and a tool for dialogue with users. The second case is *KitandPack*, a technology-based start-up which helps makers find all the components for their projects, especially in electronics. Born in the 1980s, the Maker Movement gathers creators, inventors, tinkerers, etc., who would rather build something by themselves than buy already-made ones. This movement is related to all kinds of hobbies such as cooking, electronics or woodworking. KitandPack helps makers find all the components for their projects, especially in electronics. It's a challenge: erase every kind of barrier that keeps makers away from following their ideas. The entrepreneurs want to create a profitable business in an environment led by a free exchange of knowledge and open source and where the support of a community of users and fans is crucial. Another student decided to continue the social entrepreneurship project he had developed from October to January and to implement a farmers' cooperative in a small village in Burkina Faso.

Other students have interned at very young start-ups, playing the role of a quasi-founder:

- In 2015, one student joined a start-up, DNA Script, specializing in DNA synthesis created a few months ago by a young engineer and a young Ph.D. graduate. His role was to engineer the microfluidic device in which a DNA synthesis process would take place. He was considered a co-founder of the start-up.
- The same year, another student who had developed a start-up project on digitalization for healthcare systems joined a start-up created some months ago in the same domain by serial entrepreneurs.

The majority of the students joined start-ups created 3 or 4 years ago in their development phase to take charge of some specific projects.

Lastly, some students decide to do their entrepreneurship internship in various organizations of the entrepreneurial ecosystem. In 2016, one of them joined a leading French corporate VC fund to identify worldwide innovative start-ups in the building energy efficiency domain; another joined a leading U.S. consulting firm with the mission of helping a leading Nordic countries financial service group in its digitalization with entrepreneurial methods; and another one joined a non-profit to organize a global start-up competition to promote SciTech entrepreneurship (with projects from more than 150 countries).

All these students are supervised by a corresponding manager in the organization that hosts them. Their professor meets them once every 6 weeks and has an ongoing dialogue with them. This internship is full-time: from late March to late June (and can be extended to September). In late June or late September, all these projects are the subject of a report and an oral presentation at MINES ParisTech. Audience of various backgrounds attend every presentation which lasts 30 minutes and which is followed by a debate.

The objective of the IandE option is not to create maximum start-ups. Rather, it is to equip students with an entrepreneurial mind-set that is conducive to starting a company or scaling-up an existing one. This is difficult as the "number of start-ups created" has become a key metric to evaluate the entrepreneurial "performance" of universities (along with university patents). This prompts many institutions to strongly encourage their students to create a business, which can lead to an overemphasis on quantity, as opposed to quality.

A key way to evaluate the effectiveness of this entrepreneurial specialization is to follow the alumni of the program. Of the 80 alumni of this program:

- a total of 50 percent work for start-ups or VC funds,
- 25 percent work for large and medium-sized companies,
- 20 percent work for global management consulting companies, and
- 5 percent are in a Ph.D. program.

The majority of those employed found jobs in start-ups and were not founders of these organizations. That is, they joined an *existing* entrepreneurial firm. We believe that this is a positive result for the IandE specialization, in the sense that it is important to strengthen existing start-ups, rather than merely focus on generating more start-ups. Furthermore, it should be noted that there is a high mobility between start-ups and large enterprises and consulting firms (e.g., after a few years of experience, several alumni who started their carrier in a large consulting group, created or joined a start-up).

Finally, of course, start-ups are created every year by students of MINES ParisTech that enrolled in a specialization other than IandE. Expliseat is a start-up created by students of MINES ParisTech. It produces and sells the lightest aircraft seats in the world. One of its founders enrolled in a specialization in logistics. It is important to note that this start-up benefits the network and contacts of IandE.

Young MINES ParisTech alumni also create start-ups after some years of experience. Franck Le Ouay and Romain Niccoli created CRITEO (with another engineer) after 3.5 years in their first job with Microsoft. CRITEO is a technology company that specializes in performance display advertising. At the end of 2013, CRITEO was taken public on the NASDAQ for a valuation of about $2 billion. This company is now the prize sponsor of the Contractor MINES ParisTech-Criteo, which is awarded each year to an entrepreneur who is a recent graduate of MINES ParisTech. These two companies — Expliseat and Criteo — are presented in more detail in Chapter 1.

The aim of creating this IandE specialization was to enable students to develop entrepreneurial knowledge, skills and competencies, to stimulate their imagination and their entrepreneurial spirit, to develop their ability to take responsibilities and risks and execute the process of start-up creation. A key challenge of the program was to develop teaching methods to accomplish these objectives. The end result is a program with a combination of lectures, inquiries in start-ups or in incubators, meetings with entrepreneurs, workshops with professionals, and the exploration and exploitation of entrepreneurial ideas in a start-up project work. The novelty of this program certainly lies in these mixed methods of instruction, especially the use of experiential learning.

This approach modifies our conventional view of the role of the professor. The professor is no longer just an expert with a solution to the exercise, but also an instructor who is involved in the students' projects, assisting them and teaching them how to manage their projects. This "Innovation and Entrepreneurship" training requires the students to have a high level of initiative. During this training, they have to mobilize a wide range of knowledge and resources. A comparison between projects, debates and critical discussions enables the students to assess their own strengths and weaknesses. Little by little, the image of the passive student facing the professor disappears, leaving the student with more autonomy to encourage his or her entrepreneurial capacities. This is certainly one of the main changes provided by this type of training.

Conclusion

In this chapter, we considered entrepreneurship education and its relationship to student entrepreneurship. Education helps stimulate actual entrepreneurship and the effectiveness of this activity, especially in the realm of technological entrepreneurship. The latter is especially important, given the strong emphasis in recent years on accelerating the rate of commercialization of university-based research (especially, federally funded research) and the role of the university in promoting technology-based economic development (Link *et al.*, 2015).

Since the 1980s, there has been a surge of courses and programs in entrepreneurship, and a concomitant increase in the number of full-time academics who define themselves as entrepreneurship scholars and teachers. Many studies, most of them conducted by these same entrepreneurship scholars, conclude that entrepreneurship education programs have a positive impact on the intentions of students to engage in entrepreneurship. However, there is a difference between the intention to start a venture and actually doing it. Many students intend to start a venture, and even participate in some activities associated with launching a new venture, but do not actually create a venture.

We also reviewed evidence on the content and methods used in entrepreneurship courses. Entrepreneurship has been considered an academic discipline for several decades. However, its methods and concepts have changed considerably over the last decade. The narrow view of entrepreneurship has evolved beyond start-up creation and self-employment to include broader topics, such as entrepreneurship in existing organizations, as well as developing an entrepreneurial mindset. These distinctive views about entrepreneurship have important implications for the goals of entrepreneurship education. A key objective is to provide students with the knowledge they need to establish new firms. Other goals are to foster their creativity, to develop their entrepreneurial behavior and to provide them with the capabilities to transform opportunities and ideas into economic, social, cultural or financial value (Bechard and Gregoire, 2005). Student entrepreneurship programs should also instill reality into potential entrepreneurs and indeed, for some, they may come to realize that starting a venture is not for them.

Over the last 10 years, the pedagogy of entrepreneurship has changed dramatically. Our review shows that experiential learning has become much more prevalent, in order to shift students from "intentions" to "actions". Entrepreneurship education is no longer limited to the acquisition of knowledge, but also to experimentation and experiential learning where the students deal with scarce resources, a value proposition, future customers and other stakeholders, formulating a business plan and business partners. The skills and competencies acquired by students through these formats are useful both for setting up their own business and for acting entrepreneurially in various other contexts. Such experiential learning is also central to the formal development of next-generation (next-gen) leaders in entrepreneurial family businesses including fostering the development of cognitive skills, developing emotional and social intelligence, and addressing interactions with other generations of family by bringing them into the learning context (Barbera *et al.*, 2015; Salvato *et al.*, 2015).

New methods have been employed in teaching entrepreneurship in many nations, involving a heavy emphasis on experiential learning. To illustrate this, we discussed three types of entrepreneurship courses and programs focusing on experiential learning in different contexts: the Blackstone Launchpad initiative in the U.S., the U.K. (Imperial College Business School) and France (MINES ParisTech, an engineering school). These courses and programs present diverse practices to engage students to acquire entrepreneurial competencies but also to develop their entrepreneurial spirit. We also discussed an exciting new network of externally driven entrepreneurship centers at U.S. colleges and universities such as The Blackstone LanchPad program, which is also focused on experiential learning. These three types represent different program configurations that are not mutually exclusive and that may be appropriate in different institutions with different cohorts of students and different models of approach to student entrepreneurship. Universities and entrepreneurship professors need to consider which type of approach is more appropriate or feasible for their context.

Chapter 1 showed that, over the past decade, the number of firms created by students or young alumni has increased dramatically. It seems obvious that there is a strong link between the twin phenomena of the growth of entrepreneurship courses in universities and growth in start-up creation. But how do we explain the shift that is observed between the rise of entrepreneurship education from the 1980s and the more recent growth in the number of companies created by students or young alumni?

We believe that many factors explain this trend. The first is a supply side explanation. The evidence we reviewed in this chapter suggests that the initial effects of entrepreneurial courses were small because for a long time, these courses were too disconnected from actual skills and effective methods needed to create a new firm. A second explanation is the existence of a demand side effect. In many countries, it has become much more difficult to secure employment upon graduation, while at the same time, the costs of starting a venture have fallen sharply. As a result, more students now desire to enroll in an entrepreneurship course or program and to establish a start-up firm. Moreover, in some elite universities, many recent graduates are choosing a different career path than their predecessors. For example, at MINES ParisTech during the second half of the previous decade, approximately 80 percent of the graduates chose their first job in a large firm (as they have done for more than two centuries). Since the end of the 2000s, this percentage has declined every year, reaching 50 percent by 2015. In the final chapter, we discuss how universities and governments may develop further supply-side policies to meet and sustain this student entrepreneurship movement.

References

Bae, T. J., Qian, S., Miao, C., and Fiet, J. O. 2014. The relationship between entrepreneurship education and entrepreneurial intentions: A meta-analytic review. *Entrepreneurship Theory and Practice*, 38(2), 217–254.

Barbera, F., Bernhard, F., Nacht, J., and McCann, G. 2015. The relevance of a whole-person learning approach to family business education: Concepts, evidence and implications. *Academy of Management Learning and Education*, 14(3), 322–346.

Béchard, J., and Grégoire, D. 2005. Entrepreneurship education research revisited: The case of higher education. *Academy of Management Learning and Education*, 4(1), 22–43.

Clarysse, B., and Kiefer, S. 2012. *The Smart Entrepreneur*. London: Elliott & Thompson.

Colombo, M., Piva, E., and Rossi-Lamastra, C. 2015. *Student entrepreneurs from technology-based universities: The impact of course curriculum on entrepreneurial entry*. Imperial Innovation and Entrepreneurship Conference, June 18–19, 2015, Royal Society of London.

DeTienne, D. R., and Chandler, G. N. 2004. Opportunity identification and its role in the entrepreneurial classroom: A pedagogical approach and empirical test. *Academy of Management Learning and Education*, 3(3), 242–257.

Gartner, W. B. 1988. ”Who is an Entrepreneur?” is the Wrong Question. University of Illinois at Urbana–Champaign's Academy for Entrepreneurial Leadership Historical Research Reference in Entrepreneurship. Available at SSRN: http://ssrn.com/abstract=1505236.

Hayter, C. S. 2016. Constraining entrepreneurial development: A knowledge-based view of social networks among academic entrepreneurs. *Research Policy*, 45, 475–490.

Kolvereid, L., and Moen, Ø. 1997. Entrepreneurship among business graduates: Does a major in entrepreneurship make a difference? *Journal of European Industrial Training*, 21(4), 154–160.

Lerner, J., and Malmendier, U. 2013. With a little help from my (random) friends: Success and failure in post-business school entrepreneurship. *Review of Financial Studies*, 26(10), 2411–2452.

Link, A. N., Siegel, D. S., and Wright, M. 2015. *The Chicago Handbook of University Technology Transfer and Academic Entrepreneurship*. Chicago: The University of Chicago Press.

Martin, B. C., McNally, J. J., and Kay, M. J. 2013. Examining the formation of human capital in entrepreneurship: A meta-analysis of entrepreneurship education outcomes. *Journal of Business Venturing*, 28(2), 211–224.

Mason, C., and Arshed, N. 2013. Teaching entrepreneurship to university students through experiential learning: A case study. Industry and Higher Education, 27(6), 449–463, doi:10.5367/ihe.2013.0180.

Morris, M. 1998. *Entrepreneurial Intensity: Sustainable Advantages for Individuals, Organizations, and Societies*. Greenwood Publishing Group, 170 pages.

Morris, M. and Liguori, E. 2016. Annals of Entrepreneurship Education and Pedagogy Cheltenham: Edward Elgar, 2nd edition.

Mustar, P. 2009. Technology management education: innovation and entrepreneurship at MINES ParisTech, a leading french engineering school, *Academy of Management Learning and Education*, 8(3), 418–425.

Nabi, G., Liñán, F., Fayolle, A., Krueger, N., and Walmsley, A. 2017. The impact of entrepreneurship education in higher education: A systematic review and research agenda. *Academy of Management Learning and Education*, 16(2), 277–299.

Neck, H. M., and Greene, P. G. 2011. Entrepreneurship education: Known worlds and new frontiers. *Journal of Small Business Management*, 49(1), 55–70.

Pittaway, L., and Cope, J. 2007. Entrepreneurship education: A systematic review of the evidence. *International Small Business Journal*, 25(5), 479–510.

Ratzinger, D., Greenman A., Amess K., and Mosey, S. 2015. The impact of higher education on the probabilities of reaching equity investment milestones for Internet Start Ups, paper presented at *Technology Transfer Society Conference*, Dublin, November 2015.

Rauch, A., and Hulsink, W. 2015. Putting entrepreneurship education where the intention to act lies: An investigation into the impact of entrepreneurship education on entrepreneurial behavior. *Academy of Management Learning and Education*, 14(2), 187–204.

Salvato, C., Sharma, P., and Wright, M. 2015. Learning patterns and approaches to family business education around the world — issues, insights and research agenda. *Academy of Management Learning and Education*, 14(3), 307–321.

Shane, S., and Venkataraman, S. 2000. The Promise of entrepreneurship as a field of research. *The Academy of Management Review*, 25(1), 217–226.

Singer, N. 2015. Universities Race to Nurture Start-Up Founders of the Future. *New York Times*, December 28, 2015, https://www.nytimes.com/2015/12/29/technology/universities-race-to-nurture-start-up-founders-of-the-future.html?_r=0.

Souitaris, V., Zerbinati, S., and Al-Laham, A. 2007. Do entrepreneurship programs raise entrepreneurial intention of science and engineering students? The effect of learning, inspiration and resources. *Journal of Business Venturing*, 22(4), 566–591.

Van de Ven, A. H., Polley, D. E., Garud, R., and Venkataraman, S. 1999. *The Innovation Journey*. Oxford: Oxford University Press.

Venkataraman, S. 1997. The distinctive domain of entrepreneurship research. *Advances in Entrepreneurship, Firm Emergence and Growth*, 3, 119–138.

Wadhwa, V., Freeman, R., and Rissing, B. 2008. Education and Tech Entrepreneurship, Kauffman, The Foundation of Entrepreneurship, p. 14.

Walter, S., and Block, J. 2016. Outcomes of entrepreneurship education: An institutional perspective. *Journal of Business Venturing*, 31, 216–233.

Wang, Y., and Verzat, C. 2011. Generalist or specific studies for engineering entrepreneurs? Comparison of French engineering students' trajectories in two different curricula. *Journal of Small Business and Enterprise Development*, 18(2), 366–383.

Wilson, F., Kickul, J., and Marlino, D. 2007. Gender, entrepreneurial self-efficacy, and entrepreneurial career intentions: Implications for entrepreneurship education. *Entrepreneurship Theory and Practice*, 31(3), 387–406.

Chapter 4

Financing Student Start-Ups: The Diverse Funding Landscape

Introduction

In this chapter, we elaborate the element presented in Figure 1 of Chapter 1 relating to the role of funding support in the student entrepreneurship ecosystem. An attention-grabbing headline in a leading national daily newspaper in the U.K. proclaiming "Student entrepreneurs frustrated by lack of funding" (*The Guardian*, 2012) epitomized a major entrepreneurial ecosystem challenge. This article went on to lament that while more students were interested in becoming entrepreneurs, funding for these ventures had failed to keep pace. While we have seen in Chapters 2 and 3 that the development of support models and educational delivery mechanisms enable students to gain the experiential knowledge to create ventures, obtaining additional financing typically poses challenges that go beyond the traditional resources and capabilities of professors and universities.

The funding landscape for entrepreneurial ventures is typically viewed as an escalator. Ventures are viewed as passing through various stages of development with new funders providing increasing amounts and adapting their mentoring support at each stage (Westhead *et al.*, 2011). In contrast, one of the key aspects of the development of student start-ups is the general need for smaller amounts of capital that typically fall below the radar for traditional venture capital providers or even many business angels. Also, while entrepreneurial ventures may have a pecking order for funding growth with a preference for internally generated earnings before borrowing and external equity (Vanacker and Manigart, 2010), student start-ups are typically some way off generating significant earnings and have little, if any, capacity to service borrowings. At the same time, new sources of funding for entrepreneurial ventures are emerging, particularly attracting students. These sources include university seed funds, grants, business plan competition awards, successful alumni endowments to stimulate entrepreneurial ventures by students, alumni angels, crowdfunding, accelerator programs and search funds (Tables 1 and 2).

Adopting a broad pecking order approach, we discuss the nature of these funds and analyze their advantages and limitations for student start-ups. First, we review the more traditional source of funding for start-ups and their applicability to student entrepreneurship and then turn to newer forms of funding that may have resonance for these kinds of ventures.

Table 1:   Landscape of student entrepreneurship funding types and providers

Providers of funding for student entrepreneurs	Gifts/rewards/prizes	Grants	Loans	Equity
Traditional entrepreneurial finance providers	Family, fools and friends (love money) Charities and foundations	Government grants	Bank and government loans at favourable interest rates	Venture capital Business angels
New forms finance providers	Business plan competitions public and private Crowdfunding	Foundations alumni association Co-funded scholarships	Interest-free "Honor" loans by entrepreneurial networks and foundations Crowdfunding	Alumni angel funds/networks University seed capital funds Crowdfunding Accelerators Search funds

Table 2:   Number of years post-MBA for search fund

Years post-MBA	Pre-2002	2002–2007	2008–2009	2010–2011	2012–2013
Minimum	0	0	0	0	0
Median	0	1	0	1	1
Maximum	0	5	6	6	6
No MBA (%)	0	0	0	20	29
< 1 year post-MBA	100	25	67	40	29
1–3 years post-MBA	0	50	17	20	29
4–7 years post-MBA	0	25	17	20	14
8 years or more post-MBA	0	0	0	0	0

Source: Adapted from Kolarova *et al.* (2014).

This chapter will provide examples of such sources and how they have been used. In the concluding section we assess the applicability of the different sources of funds.

Traditional Finance Providers and Student Entrepreneurs

Love money

One of the main sources of financing for the majority of entrepreneurs is "love" money. This refers to (typically small) amounts of funding from what is commonly called the "3F": Friends, Family and Fools. Although statistics are not available, our experience with students at our universities indicates that these entrepreneurial projects often receive financial support from families and friends, which may also involve very informal if not perfunctory mentoring support.

Charities and foundations

Some charities also provide funding for student entrepreneurs as part of their focus on young people. For example, in the U.K., the Prince's Trust, founded by heir to the throne, Prince Charles, is the country's leading youth charity program, offering financial help and business support for entrepreneurs aged 18–30 in England, Scotland, Wales and Northern Ireland. The Trust's Enterprise program provides mentoring in the form of training, business planning and marketing tools and mentoring support to help student entrepreneurs start their own business. The Trust is a delivery partner of the government-backed Start Up Loans Company offering low-interest finance at a rate of 6.2 percent APR in amounts of up to £5,000 (*Source*: https://www.princes-trust.org.U.K./help-for-young-people/support-starting-business).

Government grants and loans at favorable interest rates

Some national governments have introduced grants to support directly or indirectly the development of student entrepreneurship. In France, for example, the Ministry of National Education, Higher Education and Research created the Prix Pépite Tremplin for Student Entrepreneurship in 2014. For the 2016 version of the program, nearly 600 start-ups created by students or who had a start-up project took part. The jury issued 53 awards to "laureates", who were students or young graduates, aged under 29, who started their business after the 1st July 2015 or who carried out an innovative start-up creation project. These laureates received a prize of €10,000 or €5,000, awarded on the actual creation of the company. Three Grand Prizes of €20,000 were awarded to the most promising projects. A total of 60 percent were still students and 40 percent were young graduates. Nearly 47 percent of the laureates were studying for a Master's degree or had a Master's degree, 10 percent had or were studying for a Ph.D. and 32 percent were pursuing engineering studies (11 percent are in other training). A total of 25 percent of the laureates were women (*Source*: http://cache.media.enseignementsup-recherche.gouv.fr/file/Mediatheque/09/0/DP_Pepite_2016_vdef_658090.pdf).

Government schemes also provide loans for start-ups at discounted rates of interest. These may not be focused solely on student entrepreneurs. For example, The Start Up Loans Company in the U.K. is a government-backed initiative, offering loans of up to £25,000 to start a business at a fixed interest rate of 6 percent per annum. The company claims 54 percent of its loans have gone to 18–30 year olds. However, while interest is at discounted rates, early stage ventures by student entrepreneurs typically are well away from generating revenues to service such bank loans.

Bank loans for student entrepreneurs

In some countries, firm creation by students has become so popular that banks have developed specific loans for this activity which avoid the problems associated with traditional bank loans. Provision of such loans is attractive for banks as a way to capture potentially profitable customers in the long term at an early stage of their careers. A notable case of such activity has occurred in France. In 2017, the Crédit Industriel et Commercial (CIC) bank created a new product: "Start étudiant entrepreneur." A €3,000 interest-free loan to help launch an entrepreneurial firm or project. CIC is a major French financial services firm, established in 1859, with approximately 25,000 employees and 3.6 million customers. What is especially interesting is that this bank advertises this loan during prime-time hours on the main popular French TV channels with a very funny advertising campaign designed to appeal to the student demographic (*Source*: https://www.youtube.com/watch?v=H4eBTxCFVAU).

Similarly, BNP Paribas, one of the largest banks in the world, created the "Student Entrepreneur Loan" to finance the first steps related to the creation of a start-up by a student. As with the CIC product, this is an interest-free loan and there are no fees. Students can borrow from €760 to €2,000, for a period of 4–36 months.

Business angels

Business angels are individuals investing their own money in new and growing privately owned ventures. Many angel investors bring not only finance to the business, but also access to business experience, strategic advice and market and customer contacts (Fraser *et al*., 2015). Angels may invest on their own, but frequently they operate as part of a group of angels, referred to as a syndicate or network. Traditionally, angels have been mostly highly experienced, white, male entrepreneurs or managers. However, our recent survey of angels indicates a substantial increase in the number of women and younger angels (Wright *et al.*, 2015). Many younger angel investors have had experience as high-tech entrepreneurs, especially in ICT sectors. Angels also play a significant role in social entrepreneurship, an area of great interest to would-be student entrepreneurs. About a quarter of angels have invested in ventures that have a social impact (Wright *et al.*, 2015).

Some angels have emerged with a particular focus on investing in student entrepreneurs. For example, Mark Hanington and James King's angel network and "seed investment" fund FIG, an acronym for "Find, Invest, Grow", combines an active network of some 250 plus angel investors with its own investment fund to back young entrepreneurs. King is reported as saying that the motivation for targeting this market was that "we knew plenty of students wanted to start companies, but wondered if anyone had supported student entrepreneurs" (Hurley, 2012). Unlike many angel networks, it does not charge fees from start-ups looking for funding, but it does take a cut of any investment. The potential benefit to student entrepreneurs, besides funding, is the access to mentorship to help shape and develop the venture.

Venture capital firms

Venture capital (VC) firms typically have minimum size of investment thresholds, which are usually out of reach for student start-ups. Indeed, VC funding seems even less likely to apply at the early stages of student start-up development than it does for faculty start-ups from universities (Lockett and Wright, 2005; Fini *et al.*, 2016a, b). Studies have shown that few start-ups even by faculty have the prospective growth that VCs seek, nor are they at the point where they are "investor ready" for VC funding (Wright *et al.*, 2006).

Student start-ups are generally less likely to involve formal mechanisms of intellectual property, such as patents. While students may need to be aware of what VCs look for in evaluating investments, student entrepreneurs struggle to see the relevance of venture capital for their proposed ventures. Instead, student entrepreneurship courses covering finance increasingly need to give pre-eminence to the other sources of funding discussed as follows.

New Forms of Finance Providers and Student Entrepreneurs

Business plan competitions

Although there is some debate about the usefulness of business plans for entrepreneurs (Honig and Karlsson, 2013; Burke *et al.*, 2010) and as we noted in Chapter 3, many universities run business plan competitions, where winners receive funding to develop their proposed ventures and may also receive some informal mentoring support. These competitions may be funded by universities themselves or corporate and other philan-thropic sponsors.

Business plan competitions exist at the university level and may be organized either by student societies, departments or university-wide structures. For example, in the U.K., the student-run Cambridge University Entrepreneurs (CUE) competition involves a £100 ideas competition, a £1k executive summary competition and a £5k business plan competition. Competitions organized by UCL Advances in U.K. include one for under-graduates and another one for post-graduates.

Adomdza (2016) cites the case of a business plan competition at Northeastern University in Boston, historically run by professors, which grew from $10,000 to $80,000 in prizes by 2006, with some 100 students and alumni submitting their ventures. The winner received a $20,000 prize and the other 9 finalists received smaller prizes. Other major business plan competitions in the U.S. are run by MIT and Rice University.

While many business plan competitions relate to one department or one university, others are run by a particular university but are open to participants nationwide. For example, the University of Georgia's Next Top Entrepreneur competition, which involves teams pitching their exist-ing business plans or business idea in front of a live audience and a panel of judges, is open to student teams from across the country, with the win-ning team being awarded $10,000.

The New York State Student Business Plan Competition is the largest such event in the U.S., based at a public university. It was launched at the University at Albany, SUNY by the Dean of the School of Business, Professor Donald Siegel, and Professor Pradeep Haldar of the SUNY Albany's College of Nanoscale Science and Engineering. It began as a regional event, focused on colleges and universities in the Capital Region of New York State, consisting of the cities of Albany, Schenectady and Troy and now has over 70 colleges and universities participating. The effort initially received financial support from Rensselaer Polytechnic Institute, the nation's oldest technological university. A key champion of the project was Michael Castellana, President of the State Employees Federated Credit Union (SEFCU) and a prominent alumnus of the University at Albany, SUNY, who provided major financial support. The competition began with a focus on renewable energy and sustainability.

In 2013, with additional financial support from the Research Foundation of the State University of New York (SUNY), the competition expanded to include the entire state of New York, with 10 regional events feeding into the state finals at the University at Albany. Over $500,000 in prize money and in-kind services were offered. At the competition, students submit business plans in four tracks: Healthcare, Clean technologies/ Sustainability, Services/Non-Profit/Social and Nanotechnology. Several of the winners of these competitions have secured substantial amounts of additional public and private investment.

A major downside of business plan competitions is oftentimes the "winner takes all" issue. Other potentially viable projects don't receive funding directly from the competition.

Interest-free loans provided by foundations and networks

In different countries, entrepreneurs can get interest-free loans from foundations, networks and charities. In France, two networks of entrepreneurs play this role of helping new entrepreneurs with financing and advice, namely Initiative France and Réseau Entreprendre. They are increasingly solicited by student entrepreneurs, for whom they are now a source of finance in the French funding landscape.

Initiative France is a network of 230 associations to support and finance French entrepreneurs. In 2015, they supported 16,000 business start-ups. Initiative France now provides specific support to student entrepreneurs. They can benefit from global support with a collaboration with the PEPITE (see Chapter 1). Before creating their venture, student entrepreneurs can be helped with the design of the project or oriented toward a relevant partner. They can also receive an interest-free loan. Once their project is ready, a volunteer committee of experts can grant the entrepreneur what is referred to as an "honor loan" (of €9,200 on average) without interest and warranty. The experience of Initiative France shows that when it grants a loan to an entrepreneurial project, this provides a certification effect, which makes banks more willing to finance the project: "1 euro of honor loan triggers on average 7 euros of bank loan". The entrepreneur who benefits from this support also gets a mentor who is a senior entrepreneur or a business executive. They help young entrepreneurs formulate a coherent strategy, deal with other challenges and provide them with contacts. But the rule is that the mentor does not interfere in management. Lastly, one of the main problems of many entrepreneurs is that they are alone, or have only their small team to talk to. Being part of the Initiative France network allows these young entrepreneurs to have contact with other entrepreneurs to discuss business challenges, as well as to integrate them into the local economic environment (*Source*: http://www.initiative-france.fr/).

Another more powerful network that provides interest-free loans to entrepreneurs who are setting up their ventures is Réseau Entreprendre. It supports new start-ups that have the potential for job creation. Companies eligible to receive support are expected to have a size similar to that of SMEs and create on average 13 jobs in 5 years. This is a network of 5,600 entrepreneurs who are volunteer coaches. Réseau Entreprendre finances an honor loan at zero rate of interest ranging from €15,000 to €45,000 (on average €30,000) without guarantee. Some ambitious student entrepreneurship projects have been financed by this network. The network also provides support, for example, a sponsor will advise and guide the entrepreneur in the creation stages for 3 years. Subsequently, collective support is available through the clubs of creators in the network. Currently 2,300 entrepreneurs are going through this support process.

Alumni angels

Alumni who have become successful entrepreneurs may be attracted back to their alma mater as adjunct professors, advisory board members, business plan competition judges, while also providing advice and coaching support to student entrepreneurs. Besides "giving back", in these roles they may also be in a position to invest in students' ventures as business angels, gaining access to potential deals which might otherwise be inaccessible.

In France, some alumni associations have seen the emergence of professional groups committed to providing young entrepreneurs with their expertise and networks. A first association created in 2004 linked the business angels of three French engineering schools (Grandes Ecoles) and was called "XMP Business Angels" for the Ecole Polytechnique (X), Ecole des Mines (M) and Ecole des Ponts et Chaussées (P). This association enlarged its membership to include many other engineering and management schools by transforming itself in 2012 into "Business Angels of the Grandes Ecoles" (BADGE for Business Angels Des Grandes Ecoles).

This association has the following four main objectives:

1. To support the creation, development and financing of innovative start-ups with high growth potential, in particular through the involvement of former students of the grandes écoles.
2. Support of these young companies in their development, provided by more experienced alumni, especially former students of the grandes écoles.
3. Linking these entrepreneurs, creators of innovative projects with high potential for development with Business Angels, including former students of the grandes écoles.
4. All actions likely to contribute to the success of the previous objectives (*Source*: www.business-angels.info).

An increasing numbers of young student entrepreneurs are invited to pitch their projects in the meetings organized by this association.

Crowdfunding

Crowdfunding has emerged recently as an important mechanism for attracting modest amounts of funding for entrepreneurial ventures (see Bruton *et al.*, 2015). Entrepreneurs seek funding by making pitches to the network of investors who have signed up with the platform. Crowdfunding platform managers typically engage, to differing degrees, in some form of screening of offerings. Once accepted, a pitch is usually live on the platform for a fixed period. Platforms typically have an all or nothing approach, meaning that crowdfunders cash in the money only if the capital pledged at the closure of the campaign ends in at least an amount equal to the funding goal.

At the end of 2011, there were 453 crowdfunding platforms globally, raising funds totaling $1.5 billion. By May 2013, there were around 1,000 platforms with estimated funds raised amounting to $5.1 billion (Massolution, 2013). The World Bank estimates that the world crowdfunding market will expand to $93 billion by 2025 (Kshetri, 2015). Studies suggest that across Europe there were around 150 crowdfunding platforms (Dushnitsky *et al.*, 2016).

There are several types of crowdfunding mechanisms. Donation crowdfunding platforms finance projects by securing small donations from a large number of donors. Reward platforms source small amounts of money from individuals in exchange for rewards. Kickstarter is one of the largest rewards-based crowdfunding platforms. Lending platforms borrow from the crowd with individuals contributing small parts of the overall loan amount. Equity crowdfunding platforms seek investment from the crowd in exchange for a share in the entrepreneur's business or project.

In contrast to other forms of crowdfunding, equity crowdfunding investments tend to be somewhat larger, with a smaller set of investors. Equity crowdfunding has grown rapidly in recent years and now accounts for 15.6 percent of total U.K. seed and venture-stage equity investment (Nesta, 2016). Individual lead investments in pitches are routinely between £100,000 and £200,000; and depending on the model that the equity crowdfunding platform follows, average investments are between

£1,000 and £3,000 (Estrin and Khavul, 2016). Even so, the minimum investments of £10 remains popular with many investors. Different models of equity crowdfunding have emerged involving nominee (Seedrs), individual (Crowdcube), syndicated shareholdings (Syndicateroom) and fund structure approaches. These different platforms introduce different roles for individual retail investors compared with more "sophisticated" angel investors and angel syndicates. Recent evidence indicates that some 45 percent of business angels are now investing alongside crowdfunding platforms and this may provide scope for the mentoring that is absent from other forms of crowdfunding (Wright *et al.*, 2015).

Some universities have created dedicated crowdfunding spaces to support student projects. As part of its curriculum, Suffolk University in Massachusetts introduced an experiential course on crowdfunding, where students launch campaigns to fund their own start-up companies through Kickstarter and Indiegogo (Suffolk University, 2016). The Ecole of Arts et Métiers in France, in partnership with Engie, a large energy company, and KissKissBankBank, a crowdfunding platform founded in France in March 2010, has created the CrAMfunding challenge. Ten start-up projects — selected from 34 — carried out by current students from the school took part in the challenge. These projects, varied from the design of a motorized supermarket trolley to a low-cost prosthetic knee, had a dedicated space on the Crowdfunding platform KissKissBankBank to find funding. The project leaders were all trained to present a project for fund-raising. They also benefitted from personalized coaching by professionals and financial support from the school to launch their crowdfunding campaigns under the best conditions. At the end of the challenge, students won prizes from the school such as the prize of the most generous donor, the highest fund-raising prize and the prize of the largest number of donors.

Other crowdfunding platforms aimed at funding student entrepreneurs have not been dedicated to one institution but provide a platform to enable alumni to invest in student projects and students to access the capital they need. For example, AlumniFunder is a platform by the alumni where they can invest in innovative projects created by students at their alma mater (Empson, 2013). Student entrepreneurs are required to register with a ".edu" email address to set them up as part of a particular collegiate and alumni network, and then post their projects in the typical crowdfunding manner of platforms such as Kickstarter. AlumniFunder oversees the degree of quality control of projects during the process and provides students with tools to help them develop and share their presentations with the alumni.

University seed funds

Some universities have also established seed capital funds to help support early stage ventures run by students. At many institutions, these funds are actively supported by the alumni. At Cass Business School in London, students benefit from a particular alumnus fund, where a wealthy alumnus gave the school £10 million to be invested in entrepreneurship businesses from the school on the condition that this fund be used to generate money. The fund was established in 2010 with the support of Peter Cullum CBE, one of Cass's most successful entrepreneurs. Professional venture capitalists run the fund, and have high expectations for potential entrepreneurs wishing to benefit from it. Professors, then dedicated mentors, help the students prepare for the meeting with the venture capitalists, to ensure that they make a good impression, as they will probably not have a second chance. The fund has had no exit yet, but expectations are high.

The student-run EBU eLab (a pseudonym) accelerator raised a Milestone Fund through charitable donations, primarily from advisory board members who were alumni and entrepreneurs, raising a total amount of $250,000 (Adomdza, 2016). Student entrepreneurs able to provide convincing start-up milestones that they aimed to achieve could obtain Milestone grants ranging from $1,000 to $10,000 to cover initial "milestone" costs such as raw materials, website, and social media development as well as prototype manufacturing cost of products and services. There was no payback obligations but there was close oversight and an expectation that, if successful, ventures would contribute to eLab in the future.

In some cases, funding structures have been established within university structures, such as at Ludwig Maximilians University (LMU) in Munich. Historically, the university did not have funding to support entrepreneurship. The approach adopted in the LMU initiative is not to attract funding from outside to invest in student start-ups but for the

Entrepreneurship Center to have a cooperation agreement with the not-for-profit German Entrepreneurship Foundation, which seeks to promote entrepreneurship in German universities, and funding raised from sponsors provided from its wholly owned subsidiary company German Entrepreneurship Gmbh, which has the right to use the LMU name. The initiative for the center was started 10 years ago and combines an academic professor in order to have credibility within the university, a philanthropist who was connected to the university as a senator and who was committed to developing entrepreneurship, and an entrepreneur, Andy Goldstein.

The initiative had a good fortune at the time since the university President and Chancellor roles were separate, with the Chancellor being very supportive. The philanthropist committed an initial €50,000 to the Foundation to support entrepreneurship at German universities. The German Entrepreneurship Gmbh was established by Andy Goldstein as a company, which provides services that generate income to support student entrepreneurship. Goldstein is not an owner of this company and does not earn a salary. The company now runs the German accelerator program in partnership with the German government. The company pays for the coaching of entrepreneurs. Sponsors go through the foundation, while service provision is negotiated directly with the company, thereby avoiding university bureaucracy. The company now generates sufficient income for the foundation which among other things supports the social entrepreneurship academy and the Global Entrepreneurship summer school. The center helps corporations find students to fill posts and in this way they see that they are creating value for sponsors rather than simply asking them for philanthropy. Neither the foundation nor the funding corporation is legally associated with the university, remaining sustainable. The center now has 12 full-time staff and a monthly budget of €50,000. LMU funds half of the office space.

Accelerators as funders

The traditional accelerator model generally offers a small amount of funding in exchange for equity. A study by Imperial College researchers Bart Clarysse, Mike Wright and Jonas van Hove showed that this ranged from an investment of £3,600–£50,000 for 3–10 percent of the equity (Clarysse *et al.*, 2015). But their study showed that there are different types of accelerators with different objectives and investment policies. Venture capital (VC)-affiliated accelerators are interested mainly in deal flow, eventually financing the most promising participants. Such private accelerators typically provide money and services in exchange for an equity stake between 5 percent and 10 percent (or an equivalent convertible note) (Miller and Bound, 2011). Accelerators associated with corporations, such as Microsoft Ventures, typically do not take equity stakes or offer funding. Instead, these accelerators add value by helping the start-ups connect with potential customers. Some accelerators, such as Healthbox, also offer some form of follow-on funding, reflecting the challenge that despite having progressed in the accelerator, the venture may still not be at a stage to be able to attract funding from traditional providers.

Academically affiliated accelerators are interested in providing relevant entrepreneurial training, with some being exclusively for students, while others include both students and faculty (Wright and Drori, 2018). Public sector agency affiliated accelerators may have more societal objectives relating to employment creation. Public sector accelerators are typically non-profit organizations that may offer stipends or modest amounts of funding but do not request any equity in return (Pauwels *et al.*, 2016).

A central part of Georgia Tech's CREATE-X accelerator initiative for students, which has helped over 40 ventures emerge as a fully launched start-up from an idea or prototype since it was introduced in 2014, is Startup Launch. This accelerator provides $20,000 in funding through an external investment fund.

As noted in Chapter 2 with the University of Aalto-based venture Startup Sauna, this accelerator supporting student start-ups does not take an equity stake. The University of Georgia Idea Accelerator is an 8-week "business boot camp" for the university's students as well as community entrepreneurs. Over the course of the program, participants are whittled down to those entrepreneurs who are ready to pursue their ideas full-time, with the winner receiving $5000.

Breznitz and Zhang (2018) examine nine accelerators across the University of Toronto campuses, part of the Campus-Linked Accelerators (CLAs) program funded by the Government of Ontario under its Youth

Jobs Strategy, forming the main support mechanisms of student entrepreneurship. The accelerators include university-wide support mechanisms, such as University of Toronto Early Stage Technology (UTEST), as well as six accelerators created by particular faculty and/or dedicated to particular disciplines, such as the Health Innovation Hub (H2i) founded by faculty members Paul Santerre and Joseph Ferenbok. The accelerators vary greatly in terms of their goals and support from increasing awareness of entrepreneurship among students to a focus on the creation of successful start-ups. Three accelerators — H2i, the Impact Centre, and UTEST — are targeted at current students or recent graduates as well as faculty members while other programs only accept current students and recent graduates. While all accelerators provide assistance in securing external finance for firms, with some involving a final pitch event where students present to a group of potential investors, Breznitz and Zhang find that firms that spend time in accelerators whose director is a habitual entrepreneur are more likely to experience product growth. More-intensive programs had a more positive effects on firms' product growth than on employment growth. Accelerators with a screening process that required all applicants to have a proof of concept showed stronger firm performance.

Search funds

Although much of the attention regarding student entrepreneurship focuses on the creation of start-ups, some students enter entrepreneurship through the acquisition of an existing business. This route is essentially a form of leveraged buyout or management buy-in which, while oftentimes associated with the use of private equity to acquire large listed corporations has a long history in the acquisition of mid-market companies by experienced entrepreneurs (Robbie and Wright, 1995, 1996). According to the renowned Center for Management Buyout Research, based at Imperial College London, which has been monitoring the market since the 1980s, in contrast to traditional acquisition by corporations, this form of acquisition involves the purchase of an existing business by an external management team, using equity and borrowings, to create an independent company with a new ownership structure (Gilligan and Wright, 2014).

In the context of student entrepreneurship, search funds may be especially important for MBA students who have prior work experience and a toolbox of analytical skills developed as part of their Master's program but who oftentimes do not have the technical expertise, or risk appetite, to create a venture in a particular domain.

A search fund is a pool of capital raised to support the efforts of an entrepreneur, or a pair of entrepreneurs, in locating and acquiring a privately held company for the purpose of operating and growing it (Luther and Rosenthal, 2014). If the initial search capital is exhausted before a target can be identified, principals may either close the fund or attempt to raise further funding to enable the search to continue. Following the acquisition of a firm, the principals of the search fund typically take on the top management position, creating value through revenue growth, efficiency improvements, use of leverage and the addition of further acquisitions. Developing a flow of potential deals can be challenging for principals with little experience of making acquisitions.

A U.S. study identified 177 search funds created between 1983 and 2013. Of the 45 new principals in the 32 new search funds raised in 2012 and 2013, almost half (49 percent) had graduated from an MBA program within a year of raising their fund (Luther and Rosenthal, 2014). The most common sectors targeted by search funds were services, internet/information technology (IT), healthcare and education. Acquisition prices of deals typically ranged between $5 million and $20 million.

A study of search funds outside the U.S. and Canada found that in the two decades from 1992 to 2013, 28 funds had been raised. Of these 28 search funds, 10 were in the United Kingdom, 5 were in Continental Europe, 6 were in Mexico, 3 were in other Latin American countries, 2 were in India, 1 was in the Middle East and 1 was in Africa. Furthermore, in the study of search funds in the U.S., the vast majority of principals (89 percent) graduated from an MBA program (some three-quarters of these from a U.S. business school), with 71 percent raising their search fund within 2 years of graduation. As of the end of 2013, 8 were reported to be either searching for an acquisition or were fund-raising for a planned acquisition, 7 had acquired and were operating a company, 5 had deviated from the search fund model, 4 had acquired and exited a business for a positive return to investors, 2 had acquired and then shut down a company at a loss to investors and 2 had ended their search without making an acquisition, resulting in the investors' loss of search capital (Kolarova *et al.*, 2014).

Stanford Graduate School of Business alumnus, Coley Andrews, co-founded Boston-based private equity firm, Pacific Lake Partners to focus exclusively on investing in search fund entrepreneurs and their companies. Typically, they support promising recent MBA graduates looking to buy and operate a B2B service businesses with enterprise value between $5 million and $30 million and which are owned by founders seeking retirement. These businesses are healthy but oftentimes need the expertise and energy that a talented recent graduate of a top MBA programs can bring. Search funds provide both finance and advice from experienced investors and former operators. Based on his experience, Andrews estimates that entrepreneurs who raise money to acquire a company have about a 75 percent chance of buying a business within 2 years; and after that they have about a 67 percent chance of success in improving the business and making a return (Johnston, 2015). Although there are risks the search fund approach may better suit those who do not want to take the risks of a start-up.

Upon graduating from Stanford in 2006, Chris Hendriksen and Andy Schoonover launched a search fund. It was difficult for them to attract investors. It was even harder to find a suitable company and to convince the owner to sell to them (Loder, 2014). They cold-called 50 companies a week and talked to 1,000 companies before they managed to find VRI, which they bought in October 2007 (Damast, 2012). Eventually they acquired VRI, a small healthcare business offering remote patient monitoring. The number of employees increased from 40 to 250 and revenues improved to over $30 million before being sold to a private equity firm.

Conclusions

Student start-ups need smaller amounts of capital than typical university-based start-ups, which are based on technologies managed by a university technology transfer office. The traditional finance escalator of the entrepreneurial finance manual is usually not relevant for this type of firm.

Some existing traditional fund providers are customizing products to fit student entrepreneurs, but new student-specific fund providers are also emerging. Rather than being steps in an escalator as firms develop, fund provision is very much a landscape of different sources aimed at the very early seed and start-up stage. These funds often but not always involve small amounts and may be restricted in terms of the amounts of follow-on finance that can be provided. Traditional venture capital or private equity seems unlikely to be relevant to student entrepreneurship except for search funds raised to enable, typically, MBA students to undertake management buy-ins.

The model we observe appears to be more akin to funding "bricolage", defined by Baker and Nelson (2005) as "making do by applying combinations of the resources at hand to new problems and opportunities". We have seen that a variety of traditional and emerging fund providers are now providing financial support for student entrepreneurship. Student entrepreneurs typically use a mix of various sources of small amounts of financing. In a classical model of enterprise financing, a lead financier brings together a syndicate of providers (Wright and Lockett, 2003). For example, a VC will typically syndicate with the same type of funds. More recently, the entrepreneurial finance landscape has seen a marked rise in co-investment between different types of providers (Wright *et al.*, 2015). In the student model, there is no longer a lead financier in this sense. Rather, the student entrepreneurs put together different sources of finance.

However, although it is oftentimes neglected, entrepreneurship can involve the purchase and reinvigoration of an existing venture and this may be an appropriate route for more experienced MBA students. In developing a diverse approach to student entrepreneurship, universities may therefore need to consider how they can attract different kinds of funding to support these different types of entrepreneurial ventures. Student entrepreneurship models developed by missionary professors and entrepreneurship clubs may be able to raise modest amounts from alumni and their business contacts to fund prizes for start-up projects. We suggest that, for more substantial and sustainable sums to be raised, universities become strategically involved in integrated student entrepreneurship support models.

References

Adomdza, G. 2016. Choosing between a student-run and professionally managed accelerator. *Entrepreneurship Theory and Practice*, 40(4), 943–956.

Baker, T., and Nelson, R. E. 2005. Creating something from nothing: Resource construction through entrepreneurial bricolage. *Administrative Science Quarterly*, 50(3), 329–366.

Breznitz, S., and Zhang, Q. 2018. Fostering the Growth of Student Start-ups from University Accelerators: An Entrepreneurial Ecosystem Perspective. University of Toronto. Working Paper.

Bruton, G., Khavul, S., Siegel, D.S., and Wright, M. 2015. New financial alternatives in seeding entrepreneurship: Microfinance, crowdfunding, and peer-to-peer innovations. *Entrepreneurship Theory and Practice*, 39, 9–16.

Burke, A., Fraser, S., and Greene, F. 2010. Multiple effects of business plans on new ventures. *Journal of Management Studies*, 47, 391–415.

Clarysse, B., Wright, M., and van Hove, J. 2015. *A Look Inside Accelerators*. London: Nesta.

Damast, A. 2012. Search Funds: An MBA Shortcut to the C-Suite. https://www.bloomberg.com/news/articles/2012-08-31/search-funds-an-mba-shortcut-to-the-c-suite [accessed March 26, 2017].

Dushnitsky, G., Guerini, M., Piva, E., and Rossi-Lamastra, C. 2016. Crowdfunding in Europe: State-of-the-art and determinants of platforms' creation across countries. *California Management Review,* 58(2), 44–71.

Empson, R. 2013. AlumniFunder Launches A Crowdfunding Platform Where Alumni Can Back Student Entrepreneurs. March 27, 2013. https://techcrunch.com/2013/03/27/alumnifunder-launches-a-crowdfunding-platform-where-alumni-can-back-student-entrepreneurs/ [accessed July 15, 2017].

Estrin, S., and Khavul, S. 2016. Equity Crowdfunding and the Socialization of Entrepreneurial Finance. LSE Working Paper.

Fini, R. Fu, K., Rasmussen, E., Mathison, M., and Wright, M. 2016a. Determinants of University Startup Quantity and Quality in Italy, Norway and the U.K. ERC Working Paper.

Fini R., Meoli A., Sobrero M., Ghiselli S., and Ferrante F. 2016b. *Student Entrepreneurship: Demographics, Competences and Obstacles.* Consorzio Interuniversitario AlmaLaurea.

Fraser, S., Bhaumik, S., and Wright, M. 2015. What do we know about the relationship between entrepreneurial finance and growth? *International Small Business Journal*, **33**, 70–88.

Gilligan, J., and Wright, M. 2014. *Private Equity Demystified*. London: ICAEW, 3rd edition.

Honig, B., and Karlsson, T. 2013. An institutional perspective on business planning activities for nascent entrepreneurs in Sweden and the U.S. *Administrative Sciences*, 3, 266–289.

Hurley, J. 2012. Angel network and fund targets student entrepreneurs. The Telegraph, February 14, 2012. http://www.telegraph.co.U.K./finance/businessclub/9081782/Angel-network-and-fund-targets-student-entrepreneurs.html [accessed July 15, 2017].

Johnston, T. 2015. What it Takes to be a Search Fund Entrepreneur: Acquiring an existing company has lower risks than starting your own business. Insights by Stanford Business, September 25. https://www.gsb.stanford.edu/insights/what-it-takes-be-search-fund-entrepreneur [accessed March 24, 2017].

Kolarova, L., Kelly, P., Dávila, A., and Johnson, R. 2014. International search funds — 2013: Selected Observations. IESE/Stamford University.

Kshetri, N. 2015. Success of crowd-based online technology in fundraising: An institutional perspective. *Journal of International Management*, 21, 100–116.

Lockett, A., and Wright, M. 2005. Resources, capabilities, risk capital and the creation of university spin-out companies. *Research Policy*, 34, 1043–1057.

Loder, V. 2014. The Search Fund Model: How to Become A 28-Year-Old CEO. Forbes, August 7. https://www.forbes.com/sites/vanessaloder/2014/08/07/the-search-fund-model-how-to-become-a-twenty-six-year-old-ceo-if-youre-willing-to-kiss-frogs/#4de857a51190 [accessed March 26, 2017].

Luther, J., and Rosenthal, S. 2014. Search funds — 2013: Selected observations. Stamford Graduate School of Business. Case E-521.

Massolution, 2013. 2013CF: *The Crowdfunding Industry Report*. Los Angeles: Crowdsourcing LLP.

Miller, P., and Bound, K. 2011. The Startup Factories. London: Nesta.

Nesta, L. 2016. *Pushing Boundaries: the 2015 U.K. Alternative Finance Industry Report*. London: Nesta. http://www.nesta.org.U.K./publications/pushing-boundaries-2015-U.K.-alternative-finance-industry-report#sthash.25nEjUTq.dpuf [accessed June 14, 2016].

Pauwels, C., Clarysse, B., Wright, M., and Van Hove, J. 2016. Understanding a new generation incubation model: The accelerator. *Technovation*, 50–51, 13–24.

Robbie, K., and Wright, 1995. Ownership and management change: Evidence from management buy-ins. *Journal of Management Studies*, July 1995.

Robbie, K., and Wright, M. 1996. *Management Buy-ins: Entrepreneurs, Active Investors and Corporate Restructuring*, Studies in Finance Series, MUP.

Suffolk University, 2016. Student Entrepreneurs Experience Crowdfunding through Unique Course. http://www.suffolk.edu/news/67923.php#.WWnz5_ Lrsbw [accessed July 15, 2017].

Vanacker, T., and Manigart, S. 2010. Pecking order and debt capacity considerations for high growth companies seeking finance, *Small Business Economics*, 35, 53–69.

Westhead, P., Wright, M., and McElwee, G. 2011. *Entrepreneurship: Perspectives and Cases*. London: FT Prentice Hall.

Wright, M. and Drori, I. (eds.), 2018. *Accelerators*. Cheltenham: Edward Elgar.

Wright, M., Hart, M., and Fu, K. 2015. *A Nation of Angels: Assessing the Impact of Business Angels*. U.K.BAA/CFE/ERC.

Wright, M., and Lockett, A. 2003. Structure and management of alliances: Syndication of venture capital investments, *Journal of Management Studies*, 40(8), 2073–2104.

Wright, M., Lockett, A., Clarysse, B., and Binks, M. 2006. University spin-out companies and venture capital. *Research Policy*, 35, 481–501.

Chapter 5

Student Start-Ups and the Changing Context of Academic Entrepreneurship

Introduction

In this chapter, we consider student start-ups in the context of the changing landscape of academic entrepreneurship, as represented by the external context and university environment elements in Figure 1 of Chapter 1. Over the past 30 years, we have witnessed a substantial rise in the commercialization of science and other forms of university technology transfer at colleges and universities in the U.S., Europe and Asia (Siegel and Wright, 2015a). Academics have long been entrepreneurial in identifying new research opportunities, creating new disciplines and research centers as well as finding innovative ways to fund their research (Wright *et al.*, 2012; Wadhwani *et al.*, 2017). However, these commercialization activities have taken center stage as "academic entrepreneurship", since they are located at or near universities and been focused on faculty. Academic entrepreneurship has changed dramatically since the days when universities began to establish technology transfer offices in the 1980s and 1990s (Lockett *et al.*, 2015).

When these activities were first developed on campuses, there was a strong emphasis on two key dimensions of university technology transfer, namely patenting and licensing (Siegel, 2006). Little attention was paid to the start-up dimension, since this would divert attention from potentially lucrative "block-bluster" patent licensing deals. There were very few entrepreneurship courses and programs on campus, so those involved in

117

the research enterprise were not well versed in entrepreneurship or well connected to the entrepreneurial community. Also, many universities have only recently integrated academic entrepreneurship into their economic development mission.

The second operational reason for an aggressive pursuit of academic entrepreneurship, even when it is not warranted, is increasing pressure on universities to generate money from private donors. This trend has been exacerbated by the declining national-level support for universities in Europe and the declining state-level support for U.S. universities. Many alumni donors have a strong interest in supporting entrepreneurship on campus, especially if it involves students. Indeed, many alumni commercialization funds for university-based technologies have been established at leading American public and private research universities (e.g., Columbia, the University of California at Berkeley, the University of California at San Diego, Cornell, Georgetown, Purdue and the University of Maryland). Many of these alumni commercialization funds are focused on student-based start-ups.

A third operational reason for pursuing academic entrepreneurship, even when it may not be effective, is the growth of funding from federal agencies to support academic entrepreneurship (e.g., the U.S. government's Small Business Innovation Research (SBIR)/Small Business Technology Transfer (STTR) Programs).

To place the rise and potential impact of student entrepreneurship in policy context, it is important to understand this rise in university technology transfer, which was the catalyst for the development of academic entrepreneurship activities at faculty level and subsequently among students and alumni. It is also important to assess the conclusions to be drawn from 35 years of experience with start-ups by faculty from universities and public research organizations in terms of the impact of the academic start-up firms themselves, as well as their impact on academics, universities and society. We begin with some key historical events.

The Rise of University Technology Transfer

In 1976, Bob Swanson, a young venture capitalist, and Herbert Boyer, Professor of Biochemistry at the University of California at San Francisco founded Genentech. In the early 1970s, Herbert Boyer and Stanley Cohen, a geneticist, pioneered a new scientific field called recombinant DNA technology. The Boyer–Cohen patent describing a technique for the manipulation of DNA launched the biotechnology industry but also popularized a new university commercialization model (see Kenney and Patton, 2011). In this model, university laboratories generate patents and professors create start-ups, in order to produce new drugs. These start-ups are funded by VCs and may eventually go public or become acquired by a large pharmaceutical company.

This was the trajectory of Genetech. In 1980, 4 years after its creation, Genentech went public on NASDAQ. It was the first Biotech firm on the stock exchange. In 1985 — 10 years after its creation — Genetech sold its first product, synthetic human insulin. In 1990, the company reached an agreement with Roche, which enabled Roche to purchase a 60 percent equity ownership stake in Genetech and the firm was ultimately acquired by Roche for $43 billion in 2009.

After the Genentech IPO in 1980, many other firms were created in this field (but also in hardware, software, PCs and networks, etc.). Policymakers, university administrators, VCs and academics believed that these IP-based firms would be a major source of growth, wealth, jobs creation and revenue, as they sought to stimulate entrepreneurial universities (Etzkowitz *et al.*, 2000). A wide range of instruments and structures were developed to foster this movement, from the Bayh–Dole Act (Grimaldi *et al.*, 2011), which explicitly favored start-ups and small enterprises, to the creation of technology transfer offices (henceforth, TTOs) on campuses, but also with an increase in federal budgets in the biotech field (see Mowery, 2001; Kenney and Mowery, 2014).

Alongside university, government and big pharma, another actor played a crucial role in this development, namely the venture capital industry. The creation of NASDAQ in 1971 together with the liberalization of pension funds rules for VC investments in 1976 brought massive inflows of capital into this sector (see Kenney, 2014).

Of course, companies were created by academics long before the late 1970s and early 1980s (Guagnini, 2017). In 1919, Conrad Schlumberger, a Physics Professor at the School of Mines in Paris, created (along with his brother Marcel, an engineer from the Ecole Centrale) a small business to apply methods from physics for the detection of metal ores and for the geological studies to explore oil. In 1926, this business became the Société de Prospection Électrique, giving birth in the 1930s, to two companies: Schlumberger and CGG (Mustar, 2014).

In the late 1970s and early 1980s in the U.S. and U.K., we witnessed the enactment of laws to facilitate the commercialization of research at universities, such as the Bayh–Dole Act. Most developed countries have now adopted Bayh–Dole Act-like laws with specific measures to encourage patenting, licensing and firms creation by academics. Science parks, TTOs and incubators have also multiplied on campuses (Wright *et al.*, 2007).

This movement came to be called "academic entrepreneurship". Academic entrepreneurship incorporates the production of patents/ licenses and the creation of academic start-up firms (henceforth, ASOs), which are based on research findings that typically have intellectual property (IP) protection. It is this phenomenon that has attracted considerable attention from policymakers, university administrators and academics.

Impact of Academic Entrepreneurship

Surprisingly, few studies have evaluated the impacts of ASOs. In this section, we will examine the existing empirical evidence on the following aspects of ASOs:

- the number of ASOs established each year in different countries;
- their performance, in terms of growth trajectory and of job creation;
- their performance compared to corporate start-ups, or other high-tech firms;
- their realization of gains through trade sale or IPO;
- their survival rate;
- gains to academic entrepreneurs;
- gains to universities and society.

The number of ASOs created each year in different countries

In the U.S., based on data from the Association of University Technology Managers (AUTM) survey, the number of start-ups generated from university-based intellectual property had grown substantially over the past 30 years. "Start-up companies" as defined in the AUTM Survey, "refer only to those companies that were dependent upon an institution's technology for initiation". The latest AUTM survey reveals that annual university start-up activity has risen from just over 628 companies in 2005 to over 1,080 university start-ups in 2017 (AUTM, 2018).

We know that many of these start-ups are concentrated at a small number of universities. O'Shea and colleagues' (2005) longitudinal study of start-up performance of U.S. universities (1980–2001) found "that 80 percent of universities spun off less than two companies in any given year of this period, despite attracting large investment from both federal and industry sources. The average value masks a highly skewed distribution in the data in which the most productive university, MIT, spawned 31 start-ups in one year alone" (cited by O'Shea and Allen, 2014). More recently, Astebro and colleagues (2012) found the same result: "Even among the 100 most active research institutions in the U.S., faculty start-ups are not very common. For the period 1996–2007 the mean number of start-ups per university was two, and the most likely outcome was zero (*Source*: AUTM 1996–2007 reports)".

The most productive universities in start-up creation (e.g., MIT, Stanford) generate large numbers of start-ups because of the vast human and financial resources they possess, their vast alumni networks of entrepreneurs can command, their location in regions that are effectively quasi-incubators (Wright *et al.*, 2007, p. 3).

What is the situation in Europe? In the United Kingdom, the HE-BCI survey collects data on formal start-up companies based on IP where the Higher Education Provider (HEP) maintains some ownership or not (two types of start-up companies are distinguished as follows: start-ups with some HEP ownership and start-ups not owned by HEPs). The United Kingdom seems to be the most productive country in Europe, in terms of start-ups creation. Analysis of the HE-BCI survey shows that between 1999–2000 and 2011–2012, on average, 195 start-ups were formed each year based on academic research IP. However, this rate declined to 150 in 2012–2013 and 147 in 2013–2014 (*Source*: Higher Education-Business and Community Interaction survey for U.K. higher education institutions, 1999–2000 to 2013–2014).

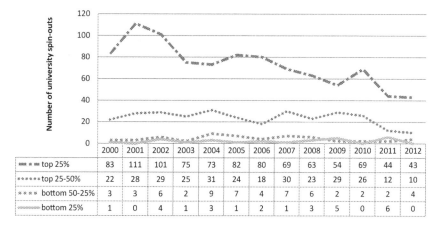

	2000	2001	2002	2003	2004	2005	2006	2007	2008	2009	2010	2011	2012
top 25%	83	111	101	75	73	82	80	69	63	54	69	44	43
top 25-50%	22	28	29	25	31	24	18	30	23	29	26	12	10
bottom 50-25%	3	3	6	2	9	7	4	7	6	2	2	2	4
bottom 25%	1	0	4	1	3	1	2	1	3	5	0	6	0

Figure 1:   Start-ups and university rankings in the U.K.

Source: Wright and Fu (2015).

In the U.K., again as in the U.S., the vast majority of start-ups are created by top quartile universities (Figure 1). However, the decline in the number of start-ups created annually is particularly notable amongst this top quartile as they shift from simply creating companies to spawning firms with the potential to add value and/or shift to licensing in cases where the IP is not suitable for a start-up (Wright and Fu, 2015).

In France, our own studies conducted between the end of the 1980s and the end of the 1990s (Mustar, 1988, 1994a, 1994b) show that at least 40 firms were created in France each year prior to the Law on Innovation and Research of 1999. From 1999 to 2008, about one hundred ASOs were created every year "in relation" to public research (Mustar, 2014). Mustar shows that the definition adopted was broader than that of the academic start-up firm, in so far as it took into account companies whose creation was based not only on the results of research work but also on projects outside research, and which formed partnerships of varying strength with a public laboratory (Mustar, 2015).

In the Netherlands, a study by van Tilburg and Kreijen (2003), described as the most comprehensive empirical survey in the Netherlands on start-up company creation by public research organizations (Zomer, 2011), estimates that 107 spin-offs were established annually by 29 public research organizations between 1999 and 2001. On average, 35 start-ups are created per year in the Netherlands from around 30 universities and research organizations.

In Spain, a study based on all 47 Spanish Public On-Campus Universities (SPOUs) between 2002 and 2006 (Rodeiro Pazso *et al.*, 2012) shows that on average 141 start-ups were created by universities per year. An average of 3.47 start-ups per university were generated on an annual basis over the time period. There is a wide dispersion between the different universities in terms of the number of start-ups they generated with one university creating more than 50 start-ups each year over the period. The study also provides interesting results in terms of the variation in start-up firms over time. There is considerable persistence in activity year on year. Over half of the universities with zero start-ups in a given year also have zero start-ups in the next analyzed year, and over 80 percent of universities with four or more start-ups in a given year also have four or more start-ups the following year.

Our own research shows that universities that have been most successful in generating the largest numbers of start-ups have clear, well-defined strategies regarding the formation and management of spin-outs (Lockett *et al.*, 2003), as well as appropriate resources and capabilities (Lockett and Wright, 2005; Mustar *et al.*, 2006). Universities generating significant numbers of start-ups typically have the most favorable policies regarding surrogate (external) entrepreneurs who have the commercial expertise that academic entrepreneurs do not. Many universities aiming to develop academic spin-offs do not invest in sufficient resources and capabilities to match their aspirations (Clarysse *et al.*, 2005), underestimating what is required to take innovations from the lab to market.

Financial and economic performance of ASOs

Many institutions have emphasized the number of start-ups created as a performance metric because that is the basis for measurement of performance of universities and technology transfer offices. Lockett *et al.* (2015) quote U.K. TTO professionals as saying "…if you were spinning-out lots of companies it made it look as though you were doing great technology transfer" and "…in the early days there was an explosion of pretty bad companies that were generally underfunded, under managed and under resourced". The raw number of start-ups created therefore says little about their economic, financial and social contributions. Many start-ups do not generate revenue from products as they are based on relatively embryonic technologies, which are far from ready for the market.

As a result, they may build value through the development of their technology (Clarysse *et al.*, 2011). For example, the biotechnology company Ablynx in Belgium was a start-up from the Flemish Biotechnology Institute (VIB) and the Vrije Universiteit Brussel. By 2000, the research group, collaborating with a multidisciplinary team at VIB had created a technology platform that could be used for the discovery and development of therapeutic drugs. The company became operational in 2002 with 10 employees. Over the next 2 years, the company raised a €5 million start-up capital from a group of international VC firms specializing in biotech and entered into collaborative and licensing agreements for the research, discovery and development of further potential drugs with leading pharmaceutical companies such as Procter and Gamble, Johnson and Johnson, Wyeth and Boehringer Ingeleheim worth several hundred million dollars.

Accordingly, analysis of performance is indicated by measures such as the extent to which they are able to attract external sources of finance, notably venture capital and business angel finance, employment growth, survival and the extent to which they have experienced a successful exit through IPO or acquisition (Siegel and Wright, 2015a, 2015b).

Access to external finance

Obtaining venture capital (VC) is often necessary to satisfy the capital requirements of start-ups. This is often central in bringing a technology from the lab to the market. Although many venture-backed firms ultimately fail, the ability to raise VC is significantly related to later success (e.g., Shane and Stuart, 2002). Moreover, VC investments constitute an important signal of the commercial potential of the university start-up by a specialist third-party expert. As such, the ability to access VC finance is one of the early stage indicators of the quality of a start-up and hence a proxy for university performance in creating quality firms.

Universities with the highest number of spin-outs obtaining venture capital finance have the most developed routines and capabilities. They use these resources and capabilities to make sure that their start-ups are ready for VC investment and networks (Lockett and Wright, 2005). The creation of a technology transfer office without this expertise is likely to be associated with an increase in the number of start-ups created, but not their quality.

Our survey of universities showed convincingly that the number of universities in a country able to attract VC funding is highly skewed (Wright *et al.*, 2006). That is due to variation in the quality of academic research conducted, the quality of the human capital on campus (faculty, postdocs and graduate students, and quality and expertise of the TTOs Start-ups created in universities located at a distance from financial centers may be at a disadvantage in raising VC finance (Smith and Ho, 2006), given that many VCs wish to physically monitor their investments. However, we also see that start-ups from universities in regions outside London and the South East are as likely to be able to raise VC if they are able to signal the quality of their ventures and the expertise of their entrepreneurs (Mueller *et al.*, 2012).

Job creation, firm growth, and survival rate

Findings across different countries are consistent in showing that the majority of academic start-ups are small, low-growth enterprises. For example, our monitoring in France showed that three-quarters were still in business 6 years after set up, but 80 percent of them had less than 10 employees (Mustar, 2001). A more recent survey (OSEO, 2013) studied 865 firms that had been in incubators linked to universities or public research organizations and that 616 firms that had won a grant in the National Competition for the creation of technologically innovative start-ups, between 1999 and 2008. In 2009, these firms totaled, respectively, 3,218 and 3,010 employees, representing an average of between 4 and 5 employees (Mustar, 2010, 2015).

Vincett (2010) shows for Canada that successful start-ups grow (often exponentially) over several decades and that their wider impacts comfortably exceed government funding of the underlying academic research.

This study's finding of a strong impact of start-ups resulting from physics research, suggests that reduced emphasis on basic work or on the basic disciplines could weaken the commercialization of academic research.

However, many start-ups in North America do not generate substantial wealth, although they do have high survival rates. A report about *The Survival of University Startups and Their Relevance to Regional Development* based on data from nine Canadian universities active in technology transfer about the start-ups they created since 1995 shows that the "survival rate" is 73 percent overall and is similar at both Medical/Doctoral institutions (72 percent) and Comprehensive institutions (76 percent) (Clayman and Holbrook, 2003). Nerkar and Shane (2003) examine a unique dataset of 128 firms founded to commercialize technologies licensed from MIT between 1980 and 1996 and also show that they have a high survival rate.

French evidence also shows the high survival rate of start-up firms. A total of 84 percent of companies supported by incubators between 1999 and 2008 were still active at the end of 2009, and 81 percent of those created just before or after winning the National Competition, in the same period, were also still active (Mustar, 2015).

Similarly, our analysis in the U.K., indicates that start-ups also seem to have a high survival rate (Figure 2). Among the university spin-out firms established between 2000 and 2012, about two-thirds (67 percent) had survived and one-third (33 percent) had failed (i.e., dissolved or in liquidation) by the July 2015 (Wright and Fu, 2015). About 79 percent of ventures which spun out in the Material/Energy/Environment sector have survived up but no less than half of firms spun out in the Commerce/ Entertainment and Management/Human Resource sector failed by this date. The failure rate was also relatively high for start-ups in Architectural/ Civil Engineering industry, at about 40 percent.

Of course, survival is only one, rather a crude measure of success. Many of these start-ups may be "living dead" (Ruhnka *et al.*, 1992) that is not technically bankrupt but not growing or generating revenues either. This is especially true in the life sciences, where the technologies are typically far removed from the market. Universities may be inefficient in liquidating these ventures if they are measured on the basis of the percentage of start-ups created that have survived. But universities may also be

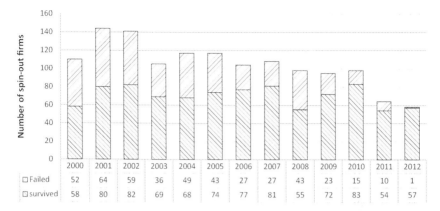

Figure 2:   U.K. start-up survival over time
Source: Wright and Fu (2015).

reluctant to liquidate start-ups where they have control, or to place pressure on academics where they only hold minority stakes, because of concerns not to offend the star scientist involved who may be very mobile to another institution that would be willing to indulge them with maintaining a non-viable start-up venture because they want the benefits of the academic's reputation (Wright *et al.*, 2007).

Science quality at the university from which the venture is spun-off and the intellectual human capital and networks of the academic entrepreneurs involved have a positive effect on start-up growth. However, the commercial orientation of research appears to have a negative effect on start-up growth (Colombo *et al.*, 2010; Nicolaou and Birley, 2003; Siegel *et al.*, 2007). Our in-depth following of start-ups' evolution after founding showed that they may be constrained by the challenges of developing competencies or capabilities that facilitate the *creation* of new development paths that depart from existing practices in the academic context (Rasmussen *et al.*, 2014). First, career academic entrepreneurs need to acquire the ability to attract new team members with industrial experience who in turn can identify and interact with industrial partners in order to develop an opportunity refinement competency. Second, career academic entrepreneurs need to evolve the credibility and entrepreneurial experience that facilitates interaction between the entrepreneurial team and external resource providers. Third, career academic entrepreneurs need to evolve a championing competency from the internal university context to include external champions. This is especially critical, due to the general lack of industrial and entrepreneurial experience among academic entrepreneurs.

The nature and extent of start-up creation and development at a particular university has been influenced by the various actors involved in academic entrepreneurship seeking to shape this activity, in order to meet their own goals, which may be at variance with those of policymakers and senior university management (Lockett *et al.*, 2015).

Comparison with other kinds of high-tech firms (corporate start-ups, etc.)

The evidence we have indicates that the performance of other forms of start-ups is better than the academic start-ups established by faculty in general. Based on a large dataset of entrepreneurial firms in Sweden, Wennberg *et al.* (2011) compared the performance of university and corporate start-ups. The authors showed that corporate start-ups (CSOs), specifically involving university graduates who had gone on to gain industrial experience, perform better than university start-ups, in terms of survival as well as growth.

Similarly, data from the U.S. Scientists and Engineers Statistical Data System (SESTAT), comparing entrepreneurship in the 1995–2006 period among faculty and university graduates with at least a bachelor's degree in science or engineering, indicate that students are a far more important source of university entrepreneurship than current or former faculty. It also appears that student companies do not seem to be of lower quality than those of current or former university employees (Astebro *et al.*, 2012).

A U.S. comparison of the performance of CSOs and start-ups from universities that had been in business for 3–10 years showed that CSOs possessed better knowledge conversion capabilities that fed through into better productivity, revenue growth and profitability than for start-ups from universities (Zahra *et al.*, 2007). A similar study we conducted of established start-ups in Belgium found that the technological knowledge transferred at start-up had an important influence on subsequent growth.

ASOs benefit most from a broad scope of technology from the parent university in term of subsequent growth, as this allows them to change market application if the first applications they pursue turn out to be a dead end, while CSOs benefit most from a specific narrow scope technology that is sufficiently distinct from the parent (Clarysse *et al.*, 2011).

ASOs also appear to have lower performance than independent new ventures, not least because they typically comprise teams that do not have sufficient spread of expertise (Ensley and Hmieleski, 2005). Start-up teams are often not sufficiently diverse in their skill sets to be able to develop the venture commercially (Vanaelst *et al.*, 2006). VCs view start-ups as being more challenging investment propositions than regular high-tech start-ups (Wright *et al.*, 2006). Besides, being more likely to require the development of a management team, start-ups typically involve more rounds of financing, a longer investment time horizon and closer monitoring. They are also more likely to be small niche market businesses.

It is useful to note that the failure rate of start-ups appears to be substantially greater than that for venture capital backed ventures (Manigart and Wright, 2013). This is not too surprising, given that start-ups are typically at an earlier life cycle stage than VC backed firms. The latter also undergo a detailed screening process prior to investment by the VC. It does suggest, however, that university TTOs may need to engage in more pre-screening before creating a start-up.

Acquisition and IPO

The exit of a start-up through IPO or strategic sale provides the opportunity for investors to realize their capital gains from academic entrepreneurship. Of the large number of start-ups created in recent years, few exit successfully through a strategic sale or IPO.

A minority of start-ups realized an exit through an IPO (26) or via a trade sale (72). The IPO percentage is much higher than in a typical population of high-tech firms where less than 5 percent make use of the IPO as an exit mechanism. Start-ups achieving an IPO are concentrated in the leading universities.

Little is known about which factors determine the probability of a trade sale or the time to trade sale. However, evidence from the U.K.

indicates that the annualized returns to start-ups that exit via a trade sale are significantly lower than for other high-tech firms (Bobelyn *et al.*, 2014).

We referred earlier to examples of successful IPOs of start-ups from universities. The uncertainties and oftentimes long timescales associated with the development of start-ups in sectors involving formal IP may mean that such companies may IPO before they have received regulatory approval let alone generated revenue from science-based products. These risks and uncertainties can mean that their longer term economic contribution may fall sharply if regulatory approval is not forthcoming. For example, the start-up firm Renovo which was developing novel anti-scarring technology, went public in London in April 2006, about 5 years after being spun-off from the University of Manchester and a little over 3 years after receiving an over-subscribed second round funding injection of $37.5 million. The following 5 years witnessed a roller-coaster ride in the company's share price as alternately good news then not so good news was announced about the progress of clinical trials of Jurista the company's primary drug (see Figure 3). After being the sixth strongest stock market performer in 2010, company fortunes declined in 2011, as Jurista failed to meet its primary or secondary endpoints. Shortly afterwards, Professor Mark Ferguson the academic scientist who had led the start-up resigned his position as CEO.

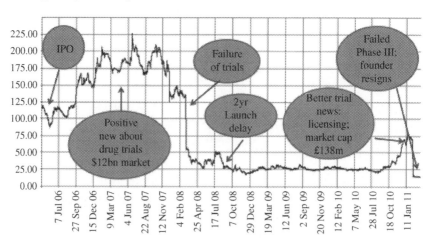

Figure 3:   Renovo share price

Gains to academics

Are academic scientists who become entrepreneurs better off than those who remain as "pure" academics? The answer appears to be a no, in financial terms, for those faculty/researchers who quit their academic positions to become full-time entrepreneurs. Evidence from academics in Sweden who quit to become an entrepreneur shows that earnings are similar before and after becoming an entrepreneur, and dividends and capital gains are inconsequential, while income risk is more than three times higher in entrepreneurship (Astebro *et al.*, 2013). Unfortunately, the authors could not assess the earnings effects for those who launch ventures but continue as academics. Comparative data on U.S. university-employed scientists with a Ph.D. in STEM disciplines leaving their university to become entrepreneurs during 1993–2006 and similar data from Sweden, Astebro *et al.* (2015) show that owning your idea outright (the "Professor's Privilege") rather than sharing ownership with your university employer (the Bayh–Dole regime) is strongly positively associated with the rate of academic entrepreneurship, but *not* with apparent economic gain for the entrepreneur (Astebro *et al.*, 2015). Additional analysis reveals that in both countries, there is too much entry into entrepreneurship, and selection from the bottom of the ability distribution among scientists.

Gains to universities and society

Academic start-ups are an international phenomenon existing in very different institutional environments. A relatively small number of IP-based start-ups or faculty start-ups are created each year in European countries and in the U.S. In Europe, as in the U.S., university and public research organization start-up activity is highly skewed. Some institutions generate many start-ups, while most others generate only a few. When we compared leading international universities and "mid-range" regionally based universities, we found marked differences (Wright *et al.*, 2008). Finally, after considerable growth in activity, especially from the late 1990s, this movement has now plateaued.

It is also important to note that officially identified start-ups substantially understate the true extent of venture creation activity by academics

(Perkmann *et al.*, 2014, 2015). Academics also create start-ups that do not depend on formal IP and which may not pass through the TTO. This has caused some debate about the dangers of IP going "out the back door" (Markman *et al.*, 2005a, b; Balven *et al.*, 2018) or academics "bypassing" the university TTO. It is not clear whether this is good or bad for society, since technologies and firms established through "bypassing" may yield significant benefits for universities and society (Shah and Pahnke, 2014). The ease with which digital companies can now be started, as compared to more traditional start-ups with formal IP, is also changing the traditional role of TTOs in the academic entrepreneurship process. The spread and successes of university-based accelerators and incubators, such as SetSquared, a consortium of five universities in southern England, is also adding to this growing trend.

Although as noted in Chapter 1, a substantial proportion of spin-offs involve doctoral students (Boh *et al.*, 2016), there also appear to be lost opportunities, in terms of converting those interested in entrepreneurship to actually creating a venture. Roach's (2017) study of early-stage doctoral students in science and engineering at leading research universities finds that entrepreneurship is widely encouraged across university research labs, ranging from 54 percent in biomedical engineering to 18 percent in particle physics, with only a small share of labs openly discouraging entrepreneurship. Such students in labs that encourage entrepreneurship do not differ from other Ph.D.s in their interest in academic careers, but they are 87 percent more likely to be interested in careers in entrepreneurship and 44 percent more likely to work in a start-up after graduation. Evidence from the Biotech Yes program in the U.K. (mainly, postdocs, see Wright *et al.*, 2012) indicates that many program participants in this scheme expressing a serious interest in starting a venture subsequently fail to do so because they join conventional companies.

We also know that the majority of research-based start-ups are quite small and remain so. While their survival rate is quite high, it is lower than the survival rate of VC-backed new ventures. Many start-ups effectively remain inactive but are not killed-off by TTOs, not least because of reluctance to upset star scientists who may threaten to move to universities more welcoming of their pet commercial projects. We also know that, in general, the performance of start-ups is lower than for other high-tech firms. Corporate start-ups perform better. Realization of gains through IPO or trade sale happen but are exceptional. Consequently, academics in general show little if any financial gain from quitting their academic job to become an entrepreneur unless they gain experience in a corporation in between (Wennberg *et al.*, 2011).

This raises the broader question of the relative impact of academic entrepreneurship activities, compared to other activities of universities. Governmental efforts to stimulate innovation, seen as a driver of the economy, brought to the fore policy decisions to promote university research that would create new industries and hence realization of the economic value of the outcomes of that research (Berman, 2015). Critics have asserted that universities seeking to develop commercialization activities are jeopardizing their fundamental mission and sacrificing the long term for short term gains (Bok, 2003). The argument is essentially that commercialization means ever greater compromises with basic academic values arising from increased secrecy in corporate-funded research, for-profit Internet companies funded by venture capitalists, industry-subsidized educational programs for physicians, conflicts of interest in research on human subjects, etc., and other questionable activities. Drawing on evidence from biology, Kleinman (2003) argues that commercial factors are having deep-rooted systematic, pervasive and indirect effects on contemporary academic practice, which may be detrimental to the public good.

However, other evidence relating basic research effort to invention disclosures suggests that pressures and mechanisms aimed at commercializing university research, such as those emanating from the Bayh–Dole Act in the U.S., have not diverted faculty from basic research to research with more commercial potential (Thursby and Thursby, 2011). Instead, both basic and applied research appear to be greater when faculty can benefit from commercialization of their research effort, in other words the two are not mutually exclusive. Similarly, Lowe and Gonzalez-Brambila (2007) analyze the research productivity of faculty members at 15 U.S. universities who formed a start-up company. These faculty members were more productive researchers than observationally equivalent colleagues before they established their firms. More importantly, the authors report that the research productivity of these academics did not decline in the aftermath of their entrepreneurial activity.

Another downside issue concerns the potential loss of faculty who leave the university to create start-ups. This is an important issue, given that academics who form start-ups tend to be star scientists. Inventors with higher perceptions of university and departmental support for entrepreneurship are less likely to leave the university when they create a start-up and more likely to maintain their affiliation (Nicolaou and Souitaris, 2016). We know from our own studies that attitudes to entrepreneurship can vary substantially between departments and disciplines even where the university is favorably disposed (Rasmussen *et al.*, 2014; Balven *et al.*, 2018). Not surprisingly, start-up entrepreneurs are less likely to leave when they have a favorable perception of departmental norms toward entrepreneurship. Universities can seek to ensure consistency between the supportive environment for entrepreneurship at institutional and department levels to avoid not only good faculty members leaving academia but potential loss of faculty champions for student entrepreneurship.

Beyond this general criticism, there has also been a major debate on how to measure the contribution of academic entrepreneurship. This debate has essentially revolved around measurement of output or measurement of activity. As we have seen, the number of start-ups created, or indeed, the number of patents or licenses, provides a very crude measure of output. Income from licenses and from the sale of shares in start-ups can be collected fairly easily and provides one hard financial measure of commercial impact. In contrast, measures of diverse activity, while feasible, may be more challenging to collect. However, while much attention is focused on licensing of patents and faculty start-up companies, these represent only a fragment of the value that industries can obtain from universities.

In the U.K., for example, the Lambert review of the knowledge transfer activities of universities argued against a narrow focus on IP and financial returns (Lambert, 2003). Indeed, most faculty do not engage in patenting (Agrawal and Henderson, 2002) or start-ups and these form only a minor part of the knowledge transferred from university labs. Further, the output and income from this activity is highly skewed (SPRU, 2002).

Evidence from the U.K. indicates that while teaching, research and entrepreneurial activities have a significant economic impact, for universities outside the top echelon, the most important activities relate to research and knowledge transfer through consulting, research contracts and research collaboration (Guerrero *et al.*, 2015). In contrast, for universities in the top echelon, entrepreneurial start-up activities have the greatest economic impact. Universities may be able to benefit financially from start-ups as a result of the indirect effect on enhanced reputation they engender (Pitsakis *et al.*, 2015). Developing a reputation for social impact via spin-offs may have positive revenue spillovers for the core university activity of research, especially for high-status universities.

Conclusions

The evidence we have presented, in this chapter, suggests that the benefits to universities and society from faculty academic entrepreneurship, especially following the introduction of Bayh–Dole Act type regulation in the U.S. and elsewhere, has not been as great as anticipated (Grimaldi *et al.*, 2011). In light of this evidence, and tightening of funding constraints, policy debates are evolving to encourage universities to make a useful contribution to wider society. In the U.K., for example, government policy places increasing emphasis upon business engagement by universities and from 2013, the five yearly Research Excellence Framework exercise evaluating research quality requires university departments to develop Impact Cases showing how their research impacts practice and policy rather than just academia.

Recent developments in academic entrepreneurship, including student entrepreneurship, therefore need to be seen as part of a movement to rethink the role of the university in society. The long-standing debate about the role of the university contrasts "education for education's sake" perspective (e.g., Newman, 1852) with a more utilitarian purpose (e.g., Smith, 1776/1999). It is, of course, a myth that academics are facing a new phenomenon of pressure to link their research work more directly to economic needs (Martin, 2012; Guagnini, 2017). Rather, it is the period of the second half of the 20th century that is anomalous. Leading research universities in Germany in the 19th century were closely linked to industry and the German model was eventually adopted by many leading research universities in the U.S., U.K. and France in the 20th century.

The establishment and growth of "polytechnics" and "land grant" universities in the U.S. and Europe, in both centuries, also strengthened the connection between universities and the industry. The establishment of institutions such as the National Science Foundation and the Cold War defense establishment, in the aftermath of World War II, and the concomitant rise of federally funded basic research at U.S. universities, may be regarded as an aberration.

Given this changing role and purpose of universities, we have reached a juncture that requires a reassessment that takes a broader view of the nature of academic entrepreneurship (Siegel and Wright, 2015a). New opportunities for academic entrepreneurship arise from the development of informal IP, beyond the focus on formal IP in patents, and the creation of new forms of entrepreneurial ventures.

These developments also emphasize the need for universities to recognize the changing perspectives of their numerous stakeholders (Clark, 1983), more of whom are becoming involved in academic entrepreneurship. These include students, a younger generation of faculty and post-doctoral fellows who are more comfortable working with the industry than the previous generation, federal agencies that support entrepreneurship programs (e.g., the U.S. government's SBIR/STTR Programs) and alumni.

In addition, other stakeholders such as research and technology transfer officials at universities, economic development officials at the university and in the region and state, surrogate entrepreneurs, managers of incubators/accelerators and science/research/technology parks, state legislatures and other bodies that govern universities have changing roles to play in creating a new ecosystem for academic entrepreneurship that places greater emphasis on teaching and student entrepreneurship.

Thus, universities are now going beyond direct transfer of technology and knowledge through ASOs to encompass indirect mechanisms to make a contribution to wider society. In particular, universities are pursuing this through the integration of technology and knowledge transfer into the curriculum and other university activities, as we have seen in previous chapters. Besides start-ups by students prior to graduation, this experience may lead indirectly to entrepreneurial actions through corporate spin-offs and start-ups by alumni.

References

Agrawal, A., and Henderson, R. 2002. Putting patents in context: Exploring knowledge transfer from MIT. *Management Science*, 48(1), 44–60.

Association of University Technology Managers (AUTM) 2018. *The AUTM Licensing Survey*, Fiscal Year 2017. AUTM, Inc.: Norwalk, CT.

Astebro, T., Braunerhjelm, P., and Broström, A. 2013. The returns to academic entrepreneurship. *Industrial and Corporate Change*, 22(1), 281–311.

Astebro, T. Braguinsky, S., Braunerhjelm, P., and Broström, A. 2015. Academic entrepreneurship: Bayh–Dole versus the "Professor's Privilege". Working Paper.

Balven, R., Fenters, V., Siegel, D. S., and Waldman, D. A. 2018. Academic entrepreneurship: The roles of identity, motivation, championing, education, work-life balance, and organizational justice. *Academy of Management Perspectives*, 32(1), 21–42.

Berman, E. P. 2015. *Creating the Market University: How Academic Science Became an Economic Engine*. Princeton: Princeton University Press.

Bobelyn, A., Clarysse, B., and Wright, M. 2014. Returns to technology commercialization and the market for firms. Imperial College Business School. Working Paper.

Boh, W. F., De-Haan, U., and Strom, R. 2016. University technology transfer through entrepreneurship: Faculty and students in spinoffs. *The Journal of Technology Transfer*, 41(4), 661–669.

Bok, D. 2003. *Universities in the Marketplace: The Commercialization of Higher Education*. Princeton University Press.

Clark, B. R. 1983. *The Higher Education System: Academic Organization in Cross-national Perspective*. Berkeley: University of California Press.

Clarysse, B., Wright, M., and Van de Velde, E. 2011. Entrepreneurial origin, technology endowments and the growth of startup companies. *Journal of Management Studies*, 48, 1420–1442.

Clarysse, B., Bruneel, J., and Wright, M. 2011. Founding resources, environment and entrepreneurial strategies to create value in high tech firms, *Strategic Entrepreneurship Journal*, 5, 137–157.

Colombo, M., D'adda, D., and Piva, E. 2010. The contribution of university research to the growth of academic start-ups: an empirical analysis. *Journal of Technology Transfer*, 35, 113–140.

Ensley, M., and Hmieleski, K. 2005. A comparative study of new venture top management team composition, dynamics and performance between university based and independent start-ups. *Research Policy*, 34, 1091–1105.

Etzkowitz, H., Webster, A., Gebhardt, C., and Terra, B. R. C., 2000. The future of the university and the university of the future: Evolution of ivory tower to entrepreneurial paradigm. *Research Policy*, 29, 313–330.

Grimaldi, R., Kenney, M., Siegel, D. S., and Wright, M., 2011. 30 years after Bayh–Dole: Reassessing academic entrepreneurship. *Research Policy* 40, 1045–1057.

Guagnini, A. 2017. Ivory towers? The commercial activity of British professors of engineering and physics, 1880–1914. *History and Technology An International Journal*, 33, 70–108.

Guerrero, M., Cunningham, J., and Urbano, D. 2015. Economic impact of entrepreneurial universities' activities: An exploratory study of the United Kingdom. *Research Policy*, 44, 748–764.

Kenney, M. 2014. Commercialization or Engagement: Which is of more significance for regional economies? In: Audretsch, D., Link, A., and Walshok, M. (eds.), *Oxford Handbook of Local Competitiveness*, Oxford: Oxford University Press.

Kenney, M., and D. Mowery. 2014. Introduction. In: Kenney, M., and Mowery, D. (eds.), *Public Universities and Regional Development: Insights from the University of California*. Stanford: Stanford University Press.

Kenney, M., and D. Patton. 2011. Does inventor ownership encourage university research-derived entrepreneurship? A six university comparison. *Research Policy*, 40(8), 1100–1112.

Kleinman, D. L. 2003. *Impure Cultures: University Biology and the World of Commerce.* Madison: University of Wisconsin Press.

Lockett, A., Wright, M., and Franklin, S. 2003. Technology transfer and universities' spin-out strategies. *Small Business Economics*, 20, 185–201.

Lockett, A., and Wright, M. 2005. Resources, capabilities, risk capital and the creation of university spin-out companies. *Research Policy*, 34(7), 1043–1057.

Lockett, A., Wright, M., and Wild, A. 2015. The institutionalization of third stream activities in U.K. higher education: The role of discourse and metrics. *British Journal of Management*, doi: 10.1111/1467-8551.12069.

Lowe, R. A., and Gonzalez-Brambila, C. 2007. Faculty entrepreneurs and research productivity. *Journal of Technology Transfer*, 32, 173–194.

Manigart, S., and Wright, M. 2013. The role of VC companies in their portfolio companies. *Foundations and Trends in Entrepreneurship*, 9(4–5), 365–570.

Markman, G. D., Gianiodis, P. T., Phan, H. P. and Balkin, D. B. 2005a. Innovation speed: Transferring university technology to market. *Research Policy*, 34, 1058–1075.

Markman, G. D., Phan, H. P., Balkin, D. B., and Gianiodis, P. T. 2005b. Entrepreneurship and university-based technology transfer. *Journal of Business Venturing*, 20(2), 241–263.

Martin, B. 2012. Are universities and university research under threat? Towards an evolutionary model of university speciation. *Cambridge Journal of Economics*, 36, 543–565.

Mowery, D. 2001. The U.S. National Innovation System after the Cold War. In: Laredo, P., and Mustar, P. (eds.), *Research and Innovation Policies in the New Global Economy: An International Comparative Analysis*, Edward Elgar: Cheltenham, U.K., pp. 15–46.

Mueller, C., Westhead, P., and Wright, M. 2012. Formal venture capital acquisition: Can entrepreneurs compensate for the spatial proximity benefits of South East England and star golden-triangle universities? *Environment and Planning A*, 44, 281–296.

Mustar, P. 1988. *Science et Innovation: Annuaire raisonné de la création d'entreprises technologiques par les chercheurs en France*, An annotated directory of technological companies created by researchers in France. Bilingual, Paris, Economica, 248 pages.

Mustar, P. 1994a. *Science et innovation. 1995. Annuaire raisonné de la création d'entreprises par les chercheurs*. Paris, Economica, 264 pages.

Mustar, P. 1994b. Organisations, technologies et marchés en création: la genèse des PME high tech. *Revue d'économie industrielle*, 67(1), 156–174.

Mustar P. 2001. *Startups from Public Research: Trends and Outlook*, STI Revue-Science, Technology, Industry, 26, OECD, pp. 165–172.

Mustar, P. 2014. Innovative entrepreneurship in France. *OECD Review of Innovation Policy: France*, pp. 227–266.

Mustar, P. 2015. Tools, rationale and economic impact of the French public policy to foster the creation of academic startup firms. Working paper presented at the Science-Based Entrepreneurship Workshop, May 7–8, University of Bologna Business School, Italy.

Newman, J. H. C. 1852. *The Idea of a University Defined and Illustrated*. London: Longmans, Green and Co.

Nicolaou, N., and Birley, S. 2003. Social networks in organizational emergence: The university spinout phenomenon. *Management Science*, 49(12), 1702–1725.

Nicolaou, N., and Souitaris, V. 2016. Can perceived support for entrepreneurship keep great faculty in the face of spinouts? *Journal of Product Innovation Management*.

O'Shea, R. P., Allen, T. J., Chevalier A., Roche F. 2005. Entrepreneurial orienta-tion, technology transfer and spinoff performance of US universities, *Research policy*, 34 (7), 994–1009.

O'Shea, R. P., and Allen, T. J. (eds.), 2014. *Building Technology Transfer in Research Universities: An Entrepreneurial Approach.* Cambridge, England: Cambridge University Press.

Perkmann, M., Fini, R., Ross J., Salter A., Silvestri C., Tartari V., 2014. Accounting for Impact at Imperial College London. A report on the activities and outputs by Imperial academics relevant for economic and social impact. Technical report, Department of Innovation and Entrepreneurship, Imperial College Business School, U.K.

Perkmann, M., Fini, R., Ross, J., Salter, A., Silvestri, C., and Tartari, V. 2015. Accounting for universities' impact: Using augmented data to measure aca-demic engagement and commercialization by academic scientists. *Research Evaluation*, forthcoming.

Pitsakis K, Souitaris, V. and Nicolaou, N. 2015. The Peripheral Halo Effect: Do Academic Spinoffs Influence Universities' Research Income? *Journal of Management Studies*, 52(3), 321–353.

Rasmussen, E., Mosey, S., and Wright, M. 2014. The influence of university departments on the evolution of entrepreneurial competencies in startup ventures. *Research Policy*, 43, 92–106.

Roach, M. 2017. Encouraging entrepreneurship in university labs: Research activi-ties, research outputs, and early doctorate careers. *PloS one*, 12(2), e0170444.

Rodero-Pados, D., Fernandez-Lopez, S., and Gonzalez, L. 2012. A resource-based view of university spin-off activity: New evidence from the Spanish case July 2012 Revista Europea de Direccion y Economia de la Empresa 21(3): 255–265.

Ruhnka, J., Feldman, H., and Dean, T. 1992. The "living dead" phenomenon in venture capital investments. Journal of Business Venturing, 7(2), 137–155.

Shah, S. K., and Pahnke, E. C. 2014. Parting the ivory curtain: Understanding how universities support a diverse set of startups. *The Journal of Technology Transfer*, 39(5), 780–792.

Siegel, D. S. 2006. *Technological Entrepreneurship: Institutions and Agents Involved in University Technology Transfer*, Cheltenham, U.K.: Edward Elgar Publishing.

Siegel, D. S., Veugelers, R., and Wright, M. 2007. University commercialization of IP: Policy implications. *Oxford Review of Economic Policy*, 23, 640–660.

Siegel, D. S., and Wright, M., 2015a. Academic entrepreneurship: Time for a rethink? *British Journal of Management*, 26(4), 582–595.

Siegel, D. S., and Wright, M., 2015b. University technology transfer offices, licensing, and start-ups. In: Link, A. N., Siegel, D., and Wright, M. (eds.), *Chicago Handbook of University Technology Transfer and Academic Entrepreneurship*. University of Chicago Press, Chicago, pp. 1–40.

Smith A. 1776/1999. *The Wealth of Nations: Books IV–V*. Penguin: Harmondsworth.

Smith, H. L., and Ho, K., 2006. Measuring the performance of Oxford University, Oxford Brookes University and the government laboratories' startup companies. *Research Policy*, 35, 1554–1568.

SPRU. 2002. *Measuring third stream activities: A final report to the Russell Group of universities.*

Thursby, J., and Thursby, M. 2011. Has the Bayh–Dole Act compromised basic research? *Research Policy*, 40(8), 1077–1083.

Vanaelst, I., Clarysse, B., Wright, M., St Jegers, *et al.* 2006. Entrepreneurial Team Development in Academic Spin-outs: An examination of team heterogeneity. *Entrepreneurship Theory and Practice*, 30(2), 249–272.

Vincett, P. S., 2010. The economic impacts of academic startup companies, and their implications for public policy. *Research Policy*, 39(6), 736–747.

Wennberg, K., Wiklund, J., and Wright, M., 2011. The effectiveness of university knowledge spillovers: Performance differences between university spinoffs and corporate spinoffs. *Research Policy* 40, 1128–1143.

Wright, M., Clarysse, B., Lockett, A., and Knockaert, M., 2008. Mid-range universities' linkages with industry: Knowledge types and the role of intermediaries. *Research Policy*, 37, 1205–1223.

Wright, M., Clarysse, B., Mustar, P., and Lockett, A. 2007. *Academic Entrepreneurship in Europe*. CheltenhamL Edward Elgar.

Wright, M., and Fu, K. 2015. University spin-outs in the U.K.: Demographics, finance and exits. ERC Working Paper.

Wright, M., Lockett, A., Clarysse, B., and Binks, M. 2006. University spin-out companies and venture capital. *Research Policy*, 35(4), 481–501.

Wright, M., Mosey, S., and Noke, H. 2012. Academic entrepreneurship and economic competitiveness: Rethinking the role of the entrepreneur. *Economics of Innovation and New Technology*, 21(5/6), 429–444.

Wadhwani, R. D., Galvez-Behar, G., Mercelis, J., and Guagnini, A. 2017. Academic entrepreneurship and institutional change in historical perspective. *Management and Organizational History*, 12, 175–198.

Zahra, S., Vandevelde, E., and Larraneta, B. 2007. Knowledge conversion capability and growth of corporate and university startups. *Industrial and Corporate Change*, 16(4), 569–608.

Chapter 6

Universities and Student Entrepreneurs: Managerial and Policy Recommendations

Introduction

In this chapter, we draw on insights from previous chapters to highlight key managerial and policy implications for universities and governments as they seek to advance student entrepreneurship on campus and surrounding regions. In the first section of the chapter, we examine managerial and policy implications for universities. Specifically, we consider the environmental, historical, cultural and resource implications of the four types of student entrepreneurship models described in Chapter 2. In the second section of the chapter, we consider the policy dimensions at various governmental levels and for corporations in terms of the institutional aspects, economic and social benefits for society, effects on corporations, design of ecosystems for student entrepreneurship, international governmental aspects and the gathering of data to assess the effectiveness of student entrepreneurship. The third section of the chapter outlines a research agenda. In the final section, we present some conclusions regarding the implications for universities in the formulation and implementation of policies and mechanisms to support student entrepreneurship.

University Policy and Management Regarding Student Entrepreneurship

University environment

The nature of the university's environment and its connection to the external environment have implications for the development of student entrepreneurship. While there may be benefits to universities and localities from creating a virtuous circle of student and graduate entrepreneurship, such approaches face the challenge of dealing with the implications of mixed or conflicting objectives. Externally integrated activities also pose issues relating to coordination between universities and the local context.

The ability to obtain funds from local or regional governments may help resolve some of the challenges in financing student start-ups identified in Chapter 4. Universities may also seek to fulfill their outreach mission by contributing to local economic development and addressing social exclusion. However, problems may be exacerbated with respect to support providers and investors from outside the university who may have purely financial goals. This can also pose potential conflicts between the goals of development (fund-raising) officers of universities seeking to raise funds from donors and student support labs.

On the other hand, the timescale and objectives of these funders may conflict with those of universities. Local or regional initiatives and funding may be closely linked to election cycles and the changing complexion of local government leadership. As a result, the objectives of local and regional government may be to demonstrate significant numbers of start-ups in local communities in a short time period but which may not be economically significant in terms of employment and revenue growth as well as sustainability. Leading research universities may be seeking to develop student entrepreneurship that is more nationally, globally or internationally recognized. There may be adverse reputational effects for universities where the failure of such activities and ventures is perceived to be the result of inadequate support from universities.

The benefits to universities and local economies and how these are achieved depend on the nature of the local context. We know that there are large variations across universities in terms of graduates' propensities to enter entrepreneurship (Daghbashyan and Hårsman, 2014; Jacob *et al.*, 2003).

For example, recent evidence from all institutions of higher education in Sweden shows that those institutions that are specialized and localized close to the major agglomeration of Stockholm are most likely to generate a high share of entrepreneurs among recent graduates, and generally are also the most likely to generate local entrepreneurs (Larson *et al.*, 2017). Swedish data suggests that 65 percent of alumni entrepreneurs start businesses within the region of the university they graduated from. Students graduating in a metropolitan area, and in a region with a strong presence of university peer entrepreneurs and family members, are much more likely than other graduates to locate their business in the region of graduation. Those who stay in a region to pursue their degrees also tend to stay after graduation to pursue their entrepreneurship there, particularly if other students make similar choices and if the region is a major urban area. The metropolitan effects are consistent with the importance of local opportunities, while the presence of peer entrepreneurs and family highlight the importance of social embeddedness.

Some universities primarily recruit students regionally, while other universities attract students from all over the country or from abroad. These geographical differences in recruitment may impact the localization of subsequent entrepreneurial activities of graduates. The share of university peers starting businesses in the locality where a particular individual graduates, reinforced by having parents nearby who are also entrepreneurs has a strong positive effects on graduates' location choice (Larson *et al.*, 2017).

The high proportion of graduates not shifting their region to start a business, compared to those who move, suggests that universities outside major urban areas may benefit from developing new strategies with local governmental agencies to maintain graduate entrepreneurs not born in the vicinity who may lack the local ties that enhance their location choice in the region. Policy efforts could include mentoring programs by peer entrepreneurs and incubator spaces seeking to facilitate start-ups by students prior to graduation (Amezcua *et al.*, 2013; Siegel and Wright, 2015). To be effective, such support may also benefit from post-graduation support provided by local governmental agencies. There may be attractions for such agencies given that the evidence presented in Chapter 1 indicates that graduate entrepreneurs are more ambitious in terms of seeking to grow their ventures than are non-graduates.

Universities outside major urban areas may find it attractive to facilitate local start-ups as these may also encourage further graduates to establish their firms in the region of graduation and thus, create a snowball effect. The isolation of some college towns in the U.S. has led some universities and state legislatures to provide special incentives (e.g., tax breaks and other direct and indirect subsidies) for entrepreneurial firms. Incentives can also include developing cultural and sports facilities in these locations away from major metropolitan areas that help attract the kind of creative individuals associated with entrepreneurship (Florida, 2002). There may also be opportunities to capture spillover employment and entrepreneurial benefits for non-graduates.

The most well-known state-level program in the U.S. to support university-based start-ups (which often involve students) is called START-UP NY (the acronym for SUNY Tax-free Areas to Revitalize and Transform Upstate NY), which was launched in 2014. This program incentivizes businesses to start, relocate to or expand in New York State through partnerships with colleges and universities. Businesses who wish to participate must locate on or near a college campus, support the academic mission of the institution and provide academic benefits for faculty, students and the institution. In return, businesses receive significant tax benefits and can leverage the valuable intellectual and physical assets of their partners in higher education.

Another important state-level initiative that often supports university-based start-ups is a "matching program" for the world's largest public sector venture capital program, the Small Business Innovation Research (SBIR)/Small Business Technology Transfer (STTR), which provides grants to small businesses conducting research for federal agencies, is another program that expands funding opportunities in the federal innovation research and development (R&D) arena. For example, the Hawaii SBIR/STTR Matching Grant program offers Hawaii companies up to 50 percent of Phase I and up to $500,000 for Phase II SBIR/STTR awards. As noted in Siegel and Wessner (2012), these start-ups often involve university faculty, current or former students and post-doctoral fellows who are more comfortable working with industry than the previous generation of scientists and engineers. According to Lanahan and Feldman (2018), these programs have been quite effective in stimulating additional R&D

investment. The authors report that there are now 13 states with such matching programs, namely Florida, Hawaii, Iowa, Kentucky, Maine, Michigan, Montana, Nebraska, North Carolina, Rhode Island, South Carolina, Virginia and Wisconsin.

In sum, while graduate start-ups in local universities potentially act as a means for regional economic development, the nature of externally integrated activities developed by universities may need to be fine-grained since policy conditions depend on whether universities are located in metropolitan areas or not, and whether graduates stay or not. As such, policies need also to consider the social embeddedness of graduate entrepreneurs, such as their region of birth and the relative rates of peer entrepreneurship in the university in question.

Also, while elite universities have worldwide reputations, they may be less connected to their local region than less prestigious universities who have built local connections over many years (Holstein *et al.*, 2018). Hence, there are some new potential challenges for elite universities that seek to "face both ways", locally and internationally, which need to be addressed. This may suggest scope for collaborations between universities in a locality or region but may not be straightforward unless the challenges relating to incentives, local prestige, etc., can be resolved.

Some universities are reaching out externally by internationalizing their support models, integrating offerings development in their institutions with demands for support from institutions in other countries. For example, the Entrepreneurship Commercialization and Innovation Center (ECIC) at the University of Adelaide in Australia has developed an overseas model based on the program in Adelaide, notably in France, Italy, South Africa and Vietnam. The Australian eChallenge program is targeted at students from any discipline and requires teams of 2–6 people to develop a business proposal for a new, previously unfunded project. Over a 12-week period, teams are offered help in the form of a series of workshops delivered in the particular country. Projects are judged by high-profile executives from the industry and government.

Some of the directions of these internationalizing efforts are perhaps quite surprising in that their direction is quite different from what has been seen in the past. Zrinka Tokic, eChallenge Program Manager in Adelaide, Australia, for example, says that a major attraction for the overseas universities in linking with Australia, especially those in Europe, is that there is a great interest in developing ventures that access the Asia-Pacific region. Professors and, crucially, mentors from Adelaide go out to the countries to teach the sessions. Students and teams are assigned mentors from Australia and in between visits meetings are conducted over Skype. Participants are given access to a specially designed online environment comprising 10 modules developed by the Center containing class notes, readings, videos and Powerpoint presentations.

Diversity of university history and culture

The historical trajectory and culture of a university represent key dimensions of the ecosystem for student entrepreneurship. Some universities have a teaching focus, while others are focused on research. Public universities, especially land-grant universities, also have a strong economic development/ community outreach mission, which complements any efforts to enhance academic entrepreneurship and student entrepreneurship.

On the other hand, private universities are typically not as engaged with their communities as public institutions. Even universities within the same strategic group, such as Research 1 universities in the U.S. or the Russell Group of leading universities in the U.K., may differ in how they view their role in promoting the development of entrepreneurial activities by faculty and students. This role may be anchored in the past and constrain how and to what extent a particular university develops an ecosystem for student entrepreneurship.

For example, Holstein *et al.* (2018) examined how the different approaches to the development of entrepreneurial activities by two research-intensive universities in the U.K. were influenced by differences in their relationship with the local region. In one case, local relations were considered historically problematic and the university saw itself as having an international perspective, while in the other case there was a closer regional link tightly associated with its civic past. Such differences have implications for how universities are able to reach-out to the local and regional community in order to participate in the building of the eco-system for student entrepreneurship. On the other hand, a university that has an international perspective, perhaps with campuses on different continents may be able to exploit these localities to help develop student entrepreneurial ventures that seek to enter international markets. This calls for an integrated approach to student entrepreneurship across campuses, which might, for example, include hands on exchange programs involving students developing international ventures.

Universities vary in terms of scale, scope, research quality, history and culture, location and local networks, resources and capabilities. As noted in Chapter 5, heterogeneity among universities has important implications for the extent and nature of start-ups by academics (Clarysse *et al.*, 2005). That is, universities may have different objectives and strategies relating to academic entrepreneurship, which will affect student entrepreneurship. For example, these institutions vary substantially, in terms of scale, scope, areas of specialization and the availability of resources. Universities with strong, world-class science, medical and engineering, as well as computer science faculties, may be able to generate different types of student entrepreneurship than those colleges or universities focused on arts, humanities and social sciences.

In previous chapters, we have described ecosystems at research universities to promote student entrepreneurship and mentioned some key programs, such as the Blackstone Launchpad program, which are centered at such institutions. But as entrepreneurship is a creative act, it may have as much if not more in common with the liberal arts (Rennie, 2008). Importantly, from the perspective of liberal arts colleges, integrating entrepreneurship into the curriculum can help students gain distinctive competencies and as such enable colleges to be more competitive in attracting students, and perhaps ultimately future alumni donations. For example, Middlebury College has reportedly been reinventing itself as a start-up incubator with annual symposiums, mentorship programs and funding competitions that has resulted in a number of commercial and social start-ups by students (Chen, 2015).

There are also examples of student entrepreneurship flourishing in the arts departments in research universities. At Arizona State University, the Herberger Institute for Design and the Arts offers a variety of enterprise and entrepreneurship programs to support the future of design and the arts by investing in student innovation and creativity and supporting entrepreneurship education and undertaking entrepreneurial activities and research. In 2006, Herberger launched the first university-based arts venture incubator at Arizona State University. The University of the Arts and Millikin University soon followed, starting incubator programs in 2008 and 2011, respectively. Nelson (2005) has shown that the Stanford music department has been quite active in promoting both technology transfer and student entrepreneurship.

But while there may be widespread interest among some arts students to become active entrepreneurs, our respondents commented that while some students in these areas are keen to be start their own venture others are resistant to the term "entrepreneurship" as they perceive it to contain negative connotations. Yet, they may be passionate about "making a difference" or "doing good" through a cause in some way that can be developed within a social entrepreneurship framework even though the term may be avoided. In seeking to promote student entrepreneurship curricula across the university, therefore, universities may need to develop differentiated approaches that appeal or key into the worldview of different student bodies.

Resources

The university context also influences the extent and nature of the resources and capabilities that universities are able to provide to support faculty and student entrepreneurship (Mustar *et al.*, 2006). Mismatches between university objectives regarding the promotion of entrepreneurship and the resources and capabilities they commit to achieving this objective are well known (Clarysse *et al.*, 2005). This potential for mismatch may need to be addressed in the design of an entrepreneurial ecosystem to support student entrepreneurship that takes account of the goals, scope and standing of any given college or university.

An important human capital resource relates to the internal university actors such as administrators involved in sponsored research and the commercialization of university research, university entrepreneurship centers and other research centers that partner with industry department chairs that champion entrepreneurship (Bercovitz and Feldman, 2008), and some faculty. Within universities, technology transfer offices (TTOs) have focused on providing support for faculty and postdocs to create spin-off ventures. There is some debate about the effectiveness of the support for entrepreneurship provided by TTOs (Siegel and Wright, 2015). Some TTOs have not traditionally seen their role to be to support student entrepreneurship, although this is beginning to change. Business schools are also starting to work with TTOs, and of course, their students are working with science and engineering faculty and students through business plan and co-working projects (Wright *et al.*, 2009).

While university career service offices have traditionally been geared to placing students with corporate employers or further education, the emergence of student entrepreneurs presents an opportunity and a challenge for them to adapt the support they provide. First, career service offices need to build connections with university entrepreneurship labs and the like in order to work with them to identify student entrepreneurs who will need support for the next phase. Second, career service offices need to build links with external accelerator programs so that they can have an appreciation of and guide potential student entrepreneurs toward this route to developing the next stage in their entrepreneurial career. Third, career service offices can build links with undergraduate and postgraduate entrepreneurship programs to identify students who do or who do not want to start their own business. Students who have followed such a program but who do not wish to start their own business may, nevertheless, have garnered skills that may be especially attractive to entrepreneurial corporate employers. Fourth, as with TTOs, recruitment of career service office personnel may need to include individuals with an entrepreneurship background, or at minimum, considerable experience in placing students with the start-up firms.

Corporations can provide resource support in terms of sponsoring awards and business plan competitions. Alumni and adjunct professors can provide a bridge between the internal and external environment through their roles as entrepreneurs in residence or as teachers of entrepreneurship and coaches of entrepreneurial creativity and business plan development.

Arizona State University, which is now America's largest university, is running an innovative entrepreneurship program, called Poder. Poder is a free, 5-week entrepreneurship program that is open to any community college student in Maricopa County. Maricopa County, with more than

4.3 million residents, is the fourth largest county in the U.S. This program is supported by Cisco, through its Silicon Valley Community Foundation. During this program, students develop personal success strategies, learn how to apply entrepreneurship skills to solve community issues and discover how to use technology to maximize their positive impact on the world. Poder students who develop an innovative Internet of Things venture or idea are eligible to compete in a Cisco Innovation Challenge for an opportunity to win seed funding of up to $5,000 to launch or grow their project.

In the U.K., for example, the Innovation and Digital Enterprise Alliance London (IDEALondon) is a collaboration involving Cisco, DC Thomson, and University College London (UCL) to promote digital innovation and entrepreneurship. UCL's role is to help with training the young entrepreneurs and to provide them mentors. Cisco is interested in firms in the field of Internet of Things and provides them technical training and financial support and equipment. DC Thomson is interested in publishing projects and help them with funding and networking opportunities.

Universities vary in terms of the financial resources they can draw on. Many of the world's leading research universities (e.g., Harvard, Stanford, Columbia, Yale and Johns Hopkins) are private universities with large endowments. These institutions not only employ leading scientists and engineers, but also have many wealthy alumni who want to support student entrepreneurship or invest in an alumni commercialization fund for student start-ups (e.g., Johns Hopkins, the University of Chicago, Yale, and NYU among others, have such funds).

University development officers charged with the task of fund-raising from alumni may need to devote greater attention to generating funds for student entrepreneurship, in addition to more traditional gifts to support buildings and endowed chairs. Some alumni may not be willing to provide a gift, but instead wish to make an investment in a student venture. Given the challenges that often arise in persuading alumni to make a cash contribution, it is critical to use alumni to develop all parts of the ecosystem. Specifically, while key performance indicators for philanthropy officers may not include attracting investors, other parts of the university, such as student entrepreneurship centers and pre-accelerators may be keen to collaborate with them.

Also, in some universities, there have been partial overlapping remits of TTOs and separate research–industry liaison and outreach offices regarding academic entrepreneurship, resulting in conflicts and duplication of efforts, but where support for student entrepreneurship may fall through the cracks between them. Increased coordination of these efforts seems warranted in order to help support student entrepreneurship through accessing sponsorships and the like but may be challenging.

Funding initiatives for student entrepreneurship, such as with corporate partners and alumni, may have a limited time horizon. A specific amount of financial support may be provided for a defined period for a specific objective. Universities may also provide matched funding alongside the external investor. An important question arises as to what happens after the external funding runs out and/or when the champion of the initiative in the university is no longer around. Universities likely have the incentive to maintain the student entrepreneurship activity as it provides an important differentiator in the market with regard to attracting students. This may especially be the case in respect of attracting foreign students who pay higher fees than domestic students and is important for universities as we detect a major interest in student entrepreneurship by foreign students.

Types of student entrepreneurship models: Challenges for universities

The activity models for student entrepreneurship that we examined in Chapter 2 face several challenges that need to be addressed if they are to be successful (Table 1).

Student-run

While student societies and clubs have grown rapidly and are now widespread and engaged in a variety of activities, as considered in Chapter 2, there are important questions concerning their sustainability both in terms of survivability and the limits to their growth, which need to be addressed. Relatedly, there is also the question of the heterogeneity of quality of student-run initiatives.

Table 1: Challenges of different policy models

Type	Challenges
Student-run	• Continuity • Scale-up • Sympathetic consideration of university management • Financing for activities
Missionary professor-run	• Scale-up • Faculty load allowance and career recognition
Across-university	• Departmental own initiatives, conflicts and buy-in • Across departments payments • Scale-up; seed funds • Links with local/regional ENT ecosystem
Integrated	• Departmental buy-in • Across departments payments • Leadership expertise • Reconciling co-existence of top-down/bottom-up

Student cohorts have limited time at the university. Thus, student entrepreneurship clubs need to have mechanisms in place to facilitate succession in leadership with the energy and drive that is needed, but also to ensure that the contacts with alumni, angels, other agents on campus and in the community are maintained. A chair of a student club at a leading U.K. university commented that he had observed that the nature and dynamism of his organization had changed from year to year, depending on which student was in charge. The same pattern has been observed in the U.S.

This model relies heavily on volunteers to organize events, contact speakers, etc. The more such clubs and societies become actively involved with helping students progress along the steps in creating their ventures, the greater the demands on volunteers' time for coaching activities. Conflicts with curriculum demands such as coursework deadlines, field visits and internships can lead to high volunteer turnover and significant limits to the amount and continuity of support that can be provided. Northeastern University Professor Gordon Adomdza (2016) recounts the case of a student-run accelerator that, while having considerable initial success in raising funds and in providing student-based coaching to support student start-ups, encountered such problems.

Major universities may have more than one student entrepreneurship club, since entrepreneurship could be viewed in some institutions as the sole domain of business or engineering schools. One large, leading university in the U.K. had three such clubs. When the administration suggested that they merge, in order to coordinate the efforts in an efficient manner, they were met with resistance. In this instance, students wanted the three clubs to compete, so that the best would emerge. Too much intervention by a university could stifle student enthusiasm for entrepreneurship.

The longevity and success of a student society for entrepreneurship may also depend on the continued support of the faculty and administrators. Interested faculty may provide a bridge to useful contacts but support of the dean may be crucial in gaining access to resources. But university professors move away or need to focus on publications for tenure and promotion. The case study written by Adomdza also included a highly successful university-wide business plan competition that had to be canceled because of a lack of faculty volunteers (Adomdza, 2016).

Another issue is that supportive deans may retire and be replaced by deans with another more interventionist agenda, which may stifle student initiative. The development of entrepreneurship centers with physical space for garages, creativity labs etc., can however provide a physical home for student clubs, such as the Oxford Foundry at Oxford University.

Student clubs may follow the development life cycle of any entrepreneurial venture. Initiatives, oftentimes introduced to fill a gap in provision by departments and universities, may need to become more professionalized to be able to scale-up and engage in significant coaching activities to help student start-ups.

Missionary professor-run

Missionary professor-run student entrepreneurship activities face significant challenges in scaling-up. Coaching students through the phases of creating a venture typically places substantial demands on faculty time. As the preparation time and content are oftentimes seen as quite different and not as intensive as regular teaching, deans may be reluctant to give the same work load allowance. Enthusiastic faculty may be willing to make a significant time commitment to launching student entrepreneurship initiatives, but lack of a full teaching load allowance can potentially have adverse career consequences. The extra effort devoted to teaching that is not fully rewarded can take time away from focus on publishing and grants.

Solutions to these challenges include providing sufficient load allowance for regular tenured faculty, but that does raise questions of budgetary costs. Alternatively, this model can develop the role of coaches as a specific career track.

Another issue to be addressed at the level of the academic department was what one respondent from a Nordic University referred to as the "civil war" between faculty approaching entrepreneurship from a traditional teaching and business plan perspective and those wanting to develop a more experiential approach. Development of a suite of offerings by different faculty may be a way to strike a balance between these different approaches.

Supporting entrepreneurship across the university

Student entrepreneurship activities that draw together students from across the university face a budgetary challenge, in the sense that is not always clear where support is coming from and how this activity will be viewed by different administrators. Different departments and school within universities may have different goals regarding entrepreneurship and may seek to construct their own departmental approach to entrepreneurship.

In U.S. and many other countries, for example, there is often a battle between engineering and business schools regarding which unit will serve as the hub for entrepreneurship. While on the face of it there may appear to be benefits of collaboration between science and engineering departments and business schools to develop entrepreneurship this seems to be limited both for academic faculty entrepreneurship (Wright *et al.*, 2009) and for student entrepreneurship. Science and engineering departments may have the technology, while business schools may have the expertise in teaching and coaching entrepreneurship. We have encountered examples of science and engineering departments being unwilling to commit parts of their budgets to university-wide entrepreneurship initiatives proposed by the business school. We have also encountered examples of engineering departments taking the initiative to develop credit bearing pathways for students pursuing attempts to create a venture, which include taking entrepreneurship courses offered by the business school but without coordination with the business school. The danger, as we saw in Chapter 2, is that these approaches typically lack the necessary coaching support to frame an entrepreneurial idea as a credible venture.

There appear to be many instances where initiatives have either not gotten off the ground or which have reverted to departmental provision because of reluctance by one department to pay another department such as an entrepreneurship center for the provision of student entrepreneurship programs or other support.

We detect real tensions between universities and departments as well as between departments that need to be reconciled. Different models are emerging that are either inside or outside university structures. Inside university structures, entrepreneurship centers dealing with student entrepreneurship may be constrained by being viewed cost centers with little discretion to generate income. The formal establishment of centers with outside funding, as with the initiatives we have seen by Blackstone, Kaufman and Coleman Foundations may be a means to address this issue. Universities need to recognize that they need student entrepreneurship centers to come into being but that they do not necessarily have to exert ownership over every accept, so long as they have some involvement in governance oversight that can ensure that the reputation of the university is not compromised.

We encountered cases where enterprise centers, student clubs, etc., were created independent of university structures because of fear that universities would kill the initiatives through bureaucracy. For examples, some initiatives by departments were resisted by universities because they were deemed to fail an inclusiveness criterion. Some successful entrepreneurship centers have been started by the university administration, such as that at the Danish Technical University. This may have been facilitated as the university is quite focused in its range of disciplines.

Internally integrated

The success of a top-down internally integrated activity model relies heavily on the strategic commitment of the university to provide finance, space and human resources. Senior university management needs to facilitate buy-in by departments. There is a need to align the goals and incentives of different departments with those of the university as a whole. Universities' TTOs also need to be part of the integrated activities.

Internally integrated activities place major demands on the appropriate leadership expertise. Leaders need to be credible to individual departments to be able to galvanize their involvement. They also need to have appropriate external networks to be able to attract the involvement of finance providers and entrepreneurs.

Internally integrated approaches also face the challenge of reconciling the co-existence of top-down and bottom-up initiatives. The development of internally integrated approaches is an evolutionary process and therefore requires universities to take a longer term strategic view to their development.

The evolution of such an approach is illustrated by the example of the University of Bologna, where efforts to stimulate student entrepreneurship began as a bottom-up initiative by professors in the Department of Management, rather than from a strategic decision by senior management in the university. Soon after these efforts began, initiatives to support student entrepreneurship were incorporated into formal practices and actions, which played a key complementary role and institutionalized the processes (Sala and Sobrero, 2018). The plan introduced identified four pillars to support student led start-ups. The first pillar focused on increasing student awareness and entrepreneurship competencies through a set of training activities with the AlmaEClub created as an open and interdisciplinary community of professors and technical and administrative staff of the University to support the spread of an entrepreneurial culture within the university.

In addition, four interest groups were established to generate new ideas and projects related to: (1) financial support and planning of business ideas; (2) alumni and entrepreneurial cross-contamination; (3) interdisciplinary training for the development of an entrepreneurial mind-set; and (4) structures for relations with the business ecosystem. The second pillar focused on the Scouting of Entrepreneurial ideas, through enhancement of the annual StartUp Day initiative, bolstered by the creation of a permanent start-up counter to support students in developing their entrepreneurial ideas. The third pillar aimed at sustaining the Pre-incubation of start-ups, with an "Entrepreneurial starting program (PdAI)" to help students selected through the StartUp Day initiative to build an entrepreneurial team and work to strengthen their business plan. The fourth pillar involved support for the structuring of new ventures through the creation of synergy among the different programs and opportunities present in the local region and the university's various initiatives aimed at promoting student entrepreneurship (the AlmaCube, CesenaLab and the Forlì Basement Club) and a set of agreements with funding and support providers in the regional, national and international ecosystem in order to create a network able to support the start-ups in their development. Importantly, this evolution was stimulated with the election of a new Rector (President) of the University who introduced the new role of Deputy Rector dedicated to entrepreneurship and relations with companies, the inclusion of Third Mission activities in the strategic plan and the reorganization of central administration offices to strengthen a new organizational structure dedicated to the Third Mission.

Governmental and Corporate Implications Levels

In this section, we use the quantitative and qualitative evidence presented in previous chapters to generate some lessons learned and key recommendations for policymakers and corporations relating to the university's role in promoting economic development through student entrepreneurship. It is important to note that our recommendations will be useful for all types of colleges and universities, including liberal arts colleges and non-research universities. Finally, this field is emerging and research is needed to better understand it. We propose several avenues for future research.

Institutional and policy environment

Drawing on recent research on contextual factors influencing entrepreneurial ecosystems (Autio *et al.*, 2014; Zahra and Wright, 2011), we conjecture that the ecosystem is influenced by the university's external environment, including the nature of local, state, regional and national public policies relating to university entrepreneurship, government objectives concerning the role of universities in society and the ownership of IP between universities and faculty/students. Country, regional and industrial contexts provide variety in their access to customers, suppliers, finance, human capital and other resources (Wright *et al.*, 2006, 2008a).

Different countries have adopted a variety of support mechanisms for academic entrepreneurship, with little evidence of convergence. For example, U.K. universities have developed diverse approaches to academic entrepreneurship within a policy context to develop faculty start-ups as part of a third stream of financing and with policy schemes generating revenue directly to universities. However, a recent shift in U.K. policy places greater emphasis on the indirect contribution of universities through developing the human capital of students that arises through their creation of entrepreneurial ventures (Wright, 2014). A recent report by Council for Science and Technology (2016) also recommended that universities do more to incorporate entrepreneurship education into formal student curricula, to complement informal and extracurricular support especially for students in mathematics, medicine and the biological and physical sciences where participation rates in entrepreneurship education were particularly low compared to students in law, social studies, creative arts and design as well as engineering and computer sciences.

In contrast, in France, policies to promote academic entrepreneurship are subsumed in a broader initiative on the part of the national government to stimulate technological entrepreneurship (Mustar and Wright, 2010). Policy in France has shifted recently from support for faculty and research staff start-ups to student start-ups. In 2014, the French government launched a national policy to foster student entrepreneurship. Its main measures are to create an *Entrepreneur Student Status* for all the students who have an entrepreneurial project during or just after their studies and to develop student entrepreneurship centers (called PEPITE in short for *Pôles étudiants pour l'innovation, le transfert et l'entrepreneuriat*) in French universities. Since 2010, each university and college has appointed an "entrepreneurship referent". Its mission is to promote entrepreneurial careers and to inform students about the existing support that can support their project. The referents work with the professional integration offices that exist in each university. Today, there are more than 300 entrepreneurship referents. This national network brings together experiences and exchanges best practices in entrepreneurship in higher education.

Economic and societal impact

Student entrepreneurship can potentially have a large economic and social impact. Start-ups play a crucial role in the productive dynamics of economies, since can serve as a mechanism for the development and commercialization of innovations, for job creation and a source of competition, which can stimulate incumbent firms (both small and large companies), a mechanism for the creation of new products and services that provide solutions to major societal challenges. For example, society needs entrepreneurs who can identify solutions for major challenges relating to healthcare, climate change, clean energy, energy conservation, food and water security, education, inclusive societies and an aging population.

The founders of the high growth start-ups formed to meet these challenges oftentimes come from leading universities, business schools and engineering schools. In France, for example, the majority founders of high growth start-ups are from only a dozen "Grandes Ecoles" in Business or Engineering (*Source*: ESCP Europe and EY, 2017). These start-ups created by students or young alumni contribute to a continuous stream of innovation that generates economic growth, through new products, new services, but also new forms of marketing and distribution, as well as new types of organizations and business models.

Effects on large companies

The growth of student entrepreneurship at universities also has important implications for large corporations. As noted earlier, many students who are exposed to such programs do not actually create a venture, nor are they employed by start-up companies. These students are often recruited by conventional corporations. For example, at MINES ParisTech in France, only 50 percent of those who take the entrepreneurship specialization create a new firm or work in a start-up after graduation. Around 30 percent choose a position in large existing firms with the remaining 20 percent joining an international consulting group such as BAIN, BCG, McKinsey, etc. At the University at Albany, SUNY business school, more than half of the entrepreneurship majors or minors were employed by the "big four" accounting firms (Ernst and Young, Deloitte, KPMG and PricewaterhouseCoopers) or by Wall Street firms or banks.

Graduates who take part in student entrepreneurship are attractive recruits for these companies because they have acquired entrepreneurial skills and an entrepreneurial mind-set that may help these firms to meet the entrepreneurial challenges increasingly faced by firms, often regardless of sector and to the job they occupy. They also have a better understanding of how to deal with entrepreneurs as clients. Large corporations are increasingly engaging in entrepreneurial activities in a variety of ways including acquiring start-ups, concluding partnerships with start-ups, creating corporate venture capital funds, incubators and accelerators, adopting an open innovation strategy, creating start-ups competitions and prizes, developing intrapreneurship, employing geeks, techies and gamers and hiring mathematicians and data scientists to help deliver the specific algorithms. This is often referred to as "corporate entrepreneurship". Many of the initiatives in place today did not exist 3–5 years ago. Excluding conventional customer–supplier relationships or equity participation, all CAC 40 corporations in France, that is the largest stock market listed firms, are now engaged with start-ups, whereas this was the case for less than one-third of them in 2010 (*Source*: http://www.davidavecgoliath.com/docs/etude-2018-DavidAvecGoliath.pdf).

To develop such strategies, these corporations need to understand how these entrepreneurial activities work, which may be quite different from traditional management approaches. Recruiting graduates who have pursued student entrepreneurship is a means of gaining a window into this world. Large corporations' support for university entrepreneurship centers also provides a way to train managers who will apply entrepreneurial methods to more traditional organizations than start-ups. In sum, educational programs and other initiatives promoting student entrepreneurship can lead to more effective corporate entrepreneurship.

Design of the student entrepreneurship ecosystem

A key challenge concerns the question of who designs the student entrepreneurship ecosystem. As we have shown, an ecosystem is a result of
various mechanisms and actors, in different contexts and evolves over
time. Although universities play a crucial role in this process, they do not
drive it. Student entrepreneurial ecosystems are co-created rather than one
institution being at the center of the process and managing it. Many stakeholders are engaged as co-creators such as students, faculty, university
managers, investors, angel networks, local authorities, start-ups and corporation. Each of these stakeholders has different objectives, norms, stan-
dards and values. Thus, many dimensions of the ecosystem go beyond
actions by universities.

This means that it is difficult to create a student entrepreneurship ecosystem from scratch or in a vacuum. Existing start-ups and entrepreneurial
projects are needed, if not there will be no users for the creation of pre-
incubators, incubators, accelerators, grants and seed money funds, etc.
University, alumni, local authorities, existing start-ups and large firms can
put in place actions and tools to accompany the process of constructing or
accelerating the development of the ecosystem.

Thus, a major challenge is: to what extent are ecosystems for student
entrepreneurship deliberately designed top-down or emerge organically
from below? The complexity and variety of the support models we have
identified in Chapter 2, and the variety of education programs and funding
sources in Chapters 3 and 4, respectively, suggest the need to develop
mechanisms for bringing together the range of different stakeholders.
Universities likely need to develop extensive and deep networks with
these stakeholders for these mechanisms to be effective. These networks
need to be able to link the different elements and levels of the ecosystem.
This raises issues concerning whether one single ecosystem is feasible for
a particular university or whether developing several sub-ecosystems
piecemeal is likely to be more feasible.

Given that a key element of the framework concerns the need for a
range of mechanisms to support the variety of entrepreneurial actors and
activities, a major challenge therefore concerns how an ecosystem is to be
constructed and how resistance is to be overcome. We have already

suggested that resistance to the word "entrepreneurship" among some may be addressed through a focus on social ventures with the objective of "doing good" rather than the purely commercial. There may also be resistance to entrepreneurial initiatives if they are viewed as being "top-down", and thus, less subject to faculty governance. There is some evidence that actors involved in academic entrepreneurship can emasculate top-down policy and strategies (Lockett *et al.*, 2015) and the challenges may be more complex in developing entrepreneurial ecosystems that promote student start-ups. There is therefore a need for policy and practice to find ways of reconciling top-down and bottom-up approaches.

Further, the developing ecosystem in universities needs to encourage and facilitate experimentation by student entrepreneurs. Students are attuned to new ideas, challenges and issues such as sustainability and climate change, and are not burdened by the legacy of a lifetime of engagement with earlier innovations and a skepticism about trying new things. They have time and energy to experiment and to be sanguine if their ideas don't work out. In developing the student entrepreneurship ecosystem, universities need to focus on enabling this experimentation. This may be in conflict with traditional teaching and learning programs but universities need to meet the challenge of finding ways to enable students to take time off to try things. This can include more flexibility for students to step out of and then rejoin programs. Another emerging option is for this experimentation to be counted as course credits, with the emphasis on assessing learning rather than whether it was successful or not.

A further challenge to the development of ecosystems for student entrepreneurship that takes account of local contexts concerns the harmonization across different regions and countries of common methods by students to identify, create and pursue opportunities. A number of structured tools, methods and concepts for the enactment of student entrepreneurship have become widely used across the world, such as the lean start-up process (Ries, 2011) and business model approaches (Pigneur and Osterwalder, 2010). Other powerful mechanisms that have spread worldwide involve community-based initiatives such as the Slush event, Global Entrepreneurship Week and Start-Up Week-end aimed at start-up entrepreneurs including students. Nearly 3,000 start-up weekends have been organized in more than 150 countries in the last year (*Source*: https://

startupweekend.org/), including that in Marrakech in February 2016, which welcomed 400 participants, of whom 83 percent were students. These events contribute (with their highly structured processes) to the development of a global entrepreneurial culture which strongly attenuates local particularities.

Although our analysis has focused primarily on colleges and universities in developed countries, we believe that our insights regarding the development of an ecosystem for student entrepreneurship can be applied to underdeveloped countries as well. For example, the importance of student entrepreneurship involving non-high tech opportunities and ventures may particularly resonate in universities in emerging economies in Asia, South America and Africa. As in the U.S. and Europe, there is a wide range of universities and student bodies on these continents, with some ranking among the leading world class institutions.

We envision that our approach generalizes to these contexts, despite the fact that national-level attitudes toward entrepreneurship vary substantially, as well as the extent of entrepreneurial activity. For mid-level universities (Wright *et al.*, 2008b), especially those in emerging economies where informal entrepreneurship is prevalent (Sutter *et al.*, 2018; Webb *et al.*, 2009), entrepreneurial opportunities may be more localized, be non-high tech and involve fewer resource requirements. Nevertheless, support for student entrepreneurship may be a means to help create ventures that have a greater socio-economic benefit. Enthusiastic professors may particularly be able to stimulate student entrepreneurship along the lines of Type 1b in our framework. Microfinance sources available in such contexts may be particularly suited to funding student entrepreneurship (Bruton *et al.*, 2015). Building of links with universities with a strong entrepreneurial education approach in developed countries may be a means to strengthen the expertise available to help develop student entrepreneurship in bottom of the pyramid countries.

International governmental implications

Significant and growing numbers of students are studying abroad at undergraduate, masters and doctoral levels. Many such students are attracted by the opportunities to pursue more experiential entrepreneurship programs that may not be available in their home country. Some of these students may return home and seek to pursue the venture ideas that they began to develop during their studies. These movements may be especially important from a policy perspective in countries with an entrepreneurship deficit. This may especially be important in emerging economies (Wright *et al.*, 2008) where besides directly increasing entrepreneurial activity there may be significant spillover benefits (Filatotchev *et al.*, 2011). In other cases, there may be an entrepreneurship drain if students stay abroad to develop their venture. Although some host governments in developed countries are tightening immigration policies knowledge worker mobility is often prioritized. While there may be increasing restrictions regarding the ability of foreign students to work in the country after graduation, there may be a more favorable attitude toward such graduates who are creating entrepreneurial ventures. At the same time, home countries, whether a developed or an emerging economy, presenting a less favorable environment for entrepreneurship may discourage students from returning to create their ventures. ICT ventures may be especially mobile in this way. Home country governments therefore may need to consider how they can develop policies that encourage such movement.

Data and monitoring

In Chapter 1, we presented data on trends in student entrepreneurship. Entrepreneurship education has been shown to have a positive effect on employment outcomes, such creativity and annual income earned (European Commission, 2012). Evidence from Denmark shows that college and university students who participated in entrepreneurship education are more likely than other students to run a business and that these businesses tend to be successful (FFE-YE, 2013).

However, this evidence is fragmented and universities and policymakers need to develop more fine-grained data collection, in order to best monitor performance in student entrepreneurship. In the U.K. for example, a recent report by the Council for Science and Technology (2016) recommended that universities needed to do more to evaluate the impact of formal and informal entrepreneurship education on students' job choices over time, including forming new ventures, in order to understand how they could target their offerings to be more effective.

Systematic longitudinal data on student-based start-ups would be extremely useful. This would also involve tracking alumni, since students may not become entrepreneurs while being enrolled or even soon after graduation (Nabi *et al.*, 2017). Rather, they may become entrepreneurs sometime later, after considerable experience in the workforce, and these may eventually be the more successful ventures (Wennberg *et al.*, 2011). Despite this, they may still benefit from student entrepreneurship programs. One aggregate level approach might be the development of a database such as the LISA database maintained by Statistics Sweden, which provides data on individual founders, including annual data on education, employment and changes in employment. Individual universities seeking to monitor their performance in student entrepreneurship can proactively engage in detailed efforts to follow graduates over time. There may also be opportunities as part of this activity to develop programs to meet the evolving entrepreneurial aspirations of alumni. We saw earlier from evidence of doctoral students that many who initially expressed a desire to become entrepreneurs do not do so. Such follow-on programs may also provide the benefit of reducing the attrition rate among those undergraduate and master's students who may have expressed an interest in starting a venture, but did not subsequently do so.

Research Implications

It is evident from our analysis in this book that a one-size-fits-all approach to student entrepreneurship ecosystem design is too simplistic. The complexity of the ecosystem poses major challenges for its effective implementation. Research is needed, therefore, that explores empirically the drivers of the variety and the effectiveness of student entrepreneurial ecosystems. In particular, we know little about how these ecosystems emerge and evolve. To what extent are they deliberately designed or emerge organically? Research is also needed that explores the interactions between these interconnected elements of the student entrepreneurship ecosystem and their co-evolution over time.

We have also seen that there is some overlap and ambiguity between different aspects of support mechanisms in the student entrepreneurial ecosystem. Research is needed to develop and clarify taxonomies of these support mechanisms.

We have noted that the nature of student entrepreneurship ecosystems may vary between contexts. However, we lack research on the relationships between supports and context. Are some contexts and supports more effective than others? What support mechanisms are more appropriate in different contexts? Heterogeneity relates to different elements of the ecosystem within a particular country, but also differences between countries (Fini *et al.*, 2016). We have discussed the emergence of student entrepreneurship in both elite universities and liberal arts colleges in developed economies and also growth of the phenomenon in emerging economies such as in Africa. Comparative research that explores different institutional contexts and how these constrain or facilitate different types of student entrepreneurship policies and programs is warranted. Similarly, research is needed to deepen insights regarding the relationships between the ranking of universities in terms of research quality as well as their differences in terms of focus, strategy, culture, connections with local communities, etc., and the extent and nature of the entrepreneurial student ventures that are created.

We also know little about the life cycles of these ecosystems and their elements. Rather than being static, like industries, they may have a life cycle. That is, they likely go through a process of emergence, establishment, maturity, change and decline (Autio *et al.*, 2018). To some extent, student entrepreneurship ecosystems appear to be in an emergence and establishment phase. To the extent that failure is observed so far, this would appear to be in terms of failures to match resources to activities and strategies. Universities' evolving responses appear to be to recognize the need to input further resources to enable their emerging goals for student entrepreneurship to be met. Following universities' student entrepreneurship strategies over time will enable future research to assess the drivers of success and failure.

Given their complexity, researchers who study these ecosystems should adopt a variety of approaches, including longitudinal process-oriented ethnographic research. Adoption of a path dependency perspective over a long period of time might also help researchers understand why particular trajectories of student entrepreneurship emerged and in some cases were maintained. Recent work on path dependencies has also explored the factors that may enable a shift from established path dependencies (Ahuja and Katila, 2004; Rasmussen *et al.*, 2011) and further research from this perspective may help in understanding how some universities have been able to shift the nature of their student entrepreneurship offering while others have not.

As we have seen, some universities have a long history of academic entrepreneurship by students and alumni, such as Stanford University (Eesley and Miller, 2012), independent of more formal efforts to establish ecosystems to develop student entrepreneurship. We lack historical studies of the development of entrepreneurial ventures by students and alumni. To what extent were these more closely related to educational programs that were outward-looking to industry than more "traditional" subject areas? Similarly, some student entrepreneurship programs date back to at least the middle of the 20th century. We have also seen that the development of student entrepreneurship programs takes several forms, from bottom-up approaches developed by entrepreneurial professors in particular departments through to cross-university initiatives. Similarly, some approaches focus on commercial entrepreneurship while others have an emphasis on entrepreneurship with social goals. Faculty based academic entrepreneurship has evolved over a long period and continues to change (Wright, 2018). Historical research approaches could help us understand how these programs evolved initially and how and why they have evolved to their current state. For example, to what extent have student entrepreneurship models relied on particular individuals and what happens when these individuals leave? Organizational resistance to change is a well-known general challenge which may frustrate or slow the introduction of new initiatives. To what extent has such organizational "stickiness" in universities created a road block to the development of student entrepreneurship models and to what extent and how have these been resolved over time?

Another key element in stimulating a student entrepreneurial ecosystem and holding its elements together concerns how the university conveys the purpose and functioning of the ecosystem. Are university senior managers able to enforce one approach for the university or should they allow decentralization? What are the consequences of different approaches? To what extent does the leadership and changes in that leadership at university and departmental levels influence the nature of student entrepreneurship in a particular university? This suggests that narrative approaches that examine both documentary evidence and how different layers of university management communicate about the ecosystem may be particularly fruitful.

Finally, although we have provided numerous examples of start-ups, our focus in this book has largely been on the ecosystems to support what we have called the student entrepreneurship movement. A major research agenda for the future concerns exploration at the individual and venture levels of the processes of venture creation and of the social, economic and financial impacts of those ventures.

Conclusions

In different countries and in different contexts, many tools have been developed by universities to foster the creation and development of faculty start-ups. In Chapter 5, we showed that in many cases, the results are poor, both in terms of direct effects (job and wealth creation for the region, an additional stream of revenue for the university (based on their investment in these firms or additional IP revenue), and indirect effects (university reputation, etc.). In Chapter 5, it is observed that policy debates are evolving that concern a potentially fundamental change in how the role of universities in making a useful contribution to society is perceived.

In Chapters 2–4, we have shown extensive and varied efforts by universities to support and promote student entrepreneurship are evolving in new and exciting ways. Student entrepreneurship can be viewed as an entrepreneurial movement that has the potential to transform colleges and universities, in the sense that it nicely complements three important trends: (1) the rise of technology commercialization and broader efforts to support entrepreneurship at research universities, which has been supported by the public and private sectors, including foundations and individual donors; (2) much stronger interest in promoting social entrepreneurship on campus and in surrounding regions and (3) the role of the university in promoting economic development.

A key issue confronting college and university presidents is whether their institutions should formulate a strategy to support student entrepreneurship and how to evaluate the effectiveness of it. There are at least two major differences between a focus on student entrepreneurship and one that is focused on venture creation by faculty. First, unlike faculty, students are demanding entrepreneurship courses and support for their projects. They also seem to be more receptive to both establishing a start-up and working for one than previous generations of students. This is due, in part, to the growth of entrepreneurship courses and programs and the establishment of entrepreneurial ecosystems on campus (e.g., incubators). For example, 10 years ago at MINES ParisTech, 80 percent of graduates were hired by large, established firms. Today, this proportion has fallen to 38 percent. The students of this engineering school now have a strong preference to work for smaller firms, especially start-ups, which provide them with more autonomy than large companies.

This is more of a bottom-up entrepreneurship strategy than a top-down strategy formulated by university administrators. This finding is also consistent with a recent report to the League of European Research Universities (LERU) on student entrepreneurship at research-intensive universities in Europe (Fyen *et al.*, 2019), which asserted that "universities need to embrace bottom-up initiatives that help to foster an entrepreneurial culture" (see Executive Summary). The LERU report also stressed the importance of interdisciplinary teams of students and project-based learning as key determinants of entrepreneurial success, a theme that also emerged from our qualitative analysis.

Second, the extent of this entrepreneurial movement of firm creation by students, for many of them during their studies, is way beyond what we have seen for faculty start-ups, and is 20 times stronger than for venture creation by faculty. In Chapter 1, we provided some evidence of this growing phenomenon in very different countries. Faced with this growing demand from their students, universities must take a stand. Should they meet this demand, and if so, how?

In Chapter 2, we proposed four different types of university student entrepreneurship models: (1a) student-run, (1b) missionary professor-run, (2) across university support and (3) internally integrated. In Chapter 3, we presented three different types of detailed student entrepreneurship programs: (1) externally driven experiential learning; (2) internally driven experiential focus on creating ventures and (3) a broader internally driven emphasis on inculcating a broad range of entrepreneurial expertise. What emerged from our analysis is that both pure bottom-up and pure top-down approaches have shortcomings if student entrepreneurship initiatives are to be sustainably led and resourced. As we have shown, bottom-up approaches may be effective in kick-starting student entrepreneurship support but are susceptible to the problems of faculty and student turn-over and motivations, and a lack of financial and other resources and capabilities to help ventures develop beyond start-up. Similarly, top-down approaches may experience conflicts between departments if they are not well integrated. Bottom-up and top-down models, as well as externally and internally driven models are not necessarily mutually exclusive. Our analysis also indicates that initiatives likely evolve over time to combine elements of each.

Regardless of which choice is made, it is important for a university to consider both the direct and indirect potential benefits of a policy to foster student entrepreneurship. These indirect benefits may be just as important as the direct benefits.

The direct benefits are mainly the financial returns to such a policy: by investing in student start-ups schools and universities may hope to generate substantial revenue. Unfortunately, such success stories are extremely rare. Another direct benefit is to meet student expectations, which are so important in the competition to recruit students. As more students become interested in firm creation, it is worth bearing in mind that they will be more inclined to enroll at a university offering them support for launching an entrepreneurial venture. That is why many colleges and universities are showcasing their student entrepreneurship programs and activities in admission tours; that is, when prospective students visit campus. For example, at the University at Albany, SUNY, the Blackstone Launchpad Center is featured in the university's admission tours.

The indirect longer term benefits can be of two different kinds. The first one concerns benefits in terms of successful alumni donations. Though that does not depend on them being entrepreneurs, in many universities, entrepreneurs are major donors. Although not all students pursuing an entrepreneurship route will be successful, many will generate significant wealth on which to base a donation. There is debate about how to measure the propensity of alumni to make donations. Alumni giving rates vary considerably across institutions. They tend to be quite high at elite private universities (e.g., 60 percent at Princeton University), although in general, they have been declining over time (The Alumni Factor, 2018). Outside the U.S., donation rates are much lower, yet are on the rise. For example, donations reached £1billion in the U.K. in 2015. Alumni giving rates also tend to be highly skewed outside the U.S. (Ross-CASE, 2018). There appears to be a trend toward targeting larger donations as the overall propensity for giving appears to be falling (Scutari, 2018). In this scenario, alumni who have become successful entrepreneurs may be particularly attractive to universities. U.S. evidence suggests that alumni with a close attachment to their alma maters and where their experience involved intellectual development (The Alumni Factor, 2018). Devoting efforts at developing strong support mechanisms to enable students to successfully realize their entrepreneurial aspirations may thus be a means to engender this affinity.

The second benefit is to signal to government and industry the relevance of entrepreneurship education programs and the dynamism of the university. This is especially important today, since almost all universities are making the case that their institutions are engines of economic development. Our analysis suggests, however, that for the student entrepreneurship movement and its support mechanisms to have this impact it needs to reach beyond the period of study to include alumni. As it takes time for the start-up to become a large-scale venture, there is a danger that without continuity in the support regime, ventures created during the period of study will simply fall off the end and potential benefits will be lost.

We hope that our analyses will provide actionable insights for student entrepreneurs, universities, policymakers, donors and other actors in the student entrepreneurship ecosystem. We believe it is important for colleges and universities to strongly support student entrepreneurship. Each of us has extensively studied technology commercialization and entrepreneurship at universities. We have also been involved in facilitating and developing entrepreneurship programs and initiatives that benefit students.

As such, we believe that it is key for development officers and university leaders to understand the full range of programs being offered at sister institutions, as well as their costs and benefits. Such a full appreciation of the breadth of these initiatives may be helpful to them when they are soliciting funds for similar programs. For example, as noted in Chapter 4, many alumni commercialization funds are partially or entirely devoted to support student-run ventures.

Student entrepreneurship provides a potentially important economic and social impact by universities beyond faculty entrepreneurship through spin-offs. While much attention regarding spin-offs by faculty focused on the hard sciences and engineering, we have shown that student entrepreneurship has a much wider application across university disciplines including arts and social sciences. While there is emerging evidence that it benefits students, universities, regions and society, the jury is still out in terms of extensive rigorous assessment. Like faculty entrepreneurship, there may be a danger that expectations run far ahead of what can be delivered unless universities commit to resourcing the ecosystems that are needed. Our hope is that the approaches and insights set out in this book will contribute to the realization of the potential of student entrepreneurship at colleges and universities.

References

Adomdza, G. 2016. Choosing between a student-run and professionally managed accelerator. *Entrepreneurship Theory and Practice*, 40(4), 943–956.

Ahuja, G., and Katila, R. 2004. Where do resources come from? The role of idiosyncratic situations. *Strategic Management Journal*, 25, 887–907.

Amezcua, A. S., Grimes, M. G., Bradley, S. W., and Wiklund, J. 2013. Organizational sponsorship and founding environments: A contingency view on the survival of business-incubated firms, 1994–2007. *Academy of Management Journal*, 56, 1628–1654.

Autio, E., Nambisan, S., Thomas, L., and Wright, M. 2018. Digital affordances, spatial affordances, and the genesis of entrepreneurial ecosystems. *Strategic Entrepreneurship Journal*, 12(1), 72–95.

Autio, E., Kenney, M., Mustar, P., Siegel, D., and Wright, M. 2014. Entrepreneurial innovation ecosystems and context, *Research Policy*, 43(7), 1097–1108.

Bercovitz, J., and Feldman, M. 2008. Academic entrepreneurs: Organizational change at the individual level. *Organization Science*, 19, 69–89.

Bruton, G., Khavul, S., Siegel, D., and Wright, M. 2015. New financial alternatives in seeding entrepreneurship: Microfinance, crowdfunding, and peer-to-peer innovations, *Entrepreneurship Theory and Practice*, 39(1), 9–26.

Chen, L. 2015. How Liberal Arts Colleges Reinvent Themselves As Startup Factories. Forbes, July 29. https://www.forbes.com/sites/liyanchen/2015/07/29/how-liberal-arts-colleges-reinvent-themselves-as-startup-factories/#3c0112d58960 [accessed September 18, 2015].

Clarysse, B., Wright, M., Lockett, A., van de Velde, E., and Vohora, A. 2005. Spinning off new ventures: A typology of facilitating services. *Journal of Business Venturing*, 20, 183–216.

Council for Science and Technology. 2016. *Strengthening Entrepreneurship Education to Boost Growth, Jobs and Productivity*. London: Council for Science and Technology.

Daghbashyan, Z., and Hårsman, B., 2014. University choice and entrepreneurship. Small Business Economics, 42, 729–746.

Eesley, C. and Miller, W. 2012. Impact: Stanford University's Economic Impact via Innovation and Entrepreneurship. Working Paper Stanford University.

European Commission, 2012. Effects and Impact of Entrepreneurship Programmes in Higher Education, Entrepreneurship Unit, European Commission.

FFE-YE. 2013. Impact of Entrepreneurship Education in Denmark 2011, 2012 and 2013, FFE-YE.

Filatotchev, I., Liu, X., Wright, M., and Lu, J. 2011. Knowledge spillovers through human mobility across national borders: Evidence from Zhongguancun Science Park in China, *Research Policy*, 40(3), 453–62.

Fini, R. Fu, K., Rasmussen, E., Mathison, M., and Wright, M. 2016. Determinants of University Startup Quantity and Quality in Italy, Norway and the U.K., ERC Working Paper.

Florida, R. 2002. The rise of the creative class. *The Washington Monthly*, May, pp. 15–25.

Fyen, W., Debackere, K., Olivares, M., Gfrörer, R., Stam, E., Mumby-Croft, B., and Keustermans, L. 2019. *Student entrepreneurship at research-intensive universities: From a peripheral activity towards a new mainstream*, advice paper no. 25, report to the League of European research Universities.

Holstein, J., Starkey, K., and Wright, M. 2018. Strategy and narrative in higher education. *Strategic Organization*, 16, 61–91.

Jacob, M., Lundqvist, M., and Hellsmark, H., 2003. Entrepreneurial transformations in the Swedish University system: The case of Chalmers University of Technology. *Research Policy,* 32, 1555–1568.

Lanahan, L., and Feldman, M. P. 2018. Approximating exogenous variation in R&D: Evidence from the Kentucky and North Carolina SBIR state match programs. *The Review of Economics and Statistics,* 100(4), 740–752.

Larson, J., Wennberg, K., Wiklund, J., and Wright, M. 2017. Location choices of graduate entrepreneurs. *Research Policy,* 46, 490–150.

Lockett, A., Wright, M., and Wild, A. 2015. The institutionalization of third stream activities in U.K. higher education: The role of discourse and metrics. *British Journal of Management,* doi: 10.1111/1467-8551.12069.

Mustar, P., Renault, M., Colombo, M., Piva, E., Fontes, M., Lockett, A., Wright, M., Clarysse, B. and Moray, N. 2006. Conceptualising the heterogeneity of research-based spin-offs: A multidimensional taxonomy. *Research Policy,* 35(2): 289–308.

Mustar, P. and Wright, M. 2010. Convergence or Path Dependency in Policies to Foster the Creation of University Spin-Off Firms? A Comparison of France and the United Kingdom, *Journal of Technology Transfer,* 35(1): 42–65.

Nabi, G., Liñán, F., Fayolle, A., Krueger, N., and Walmsley, A. 2017. The impact of entrepreneurship education in higher education: A systematic review and research agenda. *Academy of Management Learning and Education,* 16(2), 277–299.

Nelson, A. J. 2005. Cacophony or harmony? Multivocal logics and technology licensing by the Stanford University Department of Music, *Industrial and Corporate Change,* 14(1), 93–118.

Pigneur, Y., and Osterwalder, A. 2010. *The Business Model Generation.*

Rasmussen, E., Mosey, S., and Wright, M. 2011. The evolution of entrepreneurial competencies: A longitudinal study of university spin-off venture emergence. *Journal of Management Studies,* 48(6), 1314–1345.

Rennie, H. 2008. Entrepreneurship as a liberal art. *Politics and Policy,* 36, 197–215.

Ries, E. 2011. The Lean Startup: How Today's Entrepreneurs Use Continuous Innovation to Create Radically Successful Businesses.

Ross-CASE. 2018. Giving to Excellence: Generating Philanthropic Support for Higher Education. CASE. https://www.case.org/Documents/Research/Ross-CASE/Ross_CASE_1617_Report_Final.pdf [accessed October 8, 2018].

Sala, I., and Sobrero, M. 2018. Games of policy and practice: Multi-level dynamics and the role of Universities in knowledge transfer processes. University of Bologna. Working Paper.

Scutari, M. 2018. Mega-gifts Are Rising and Alumni Giving Is Shrinking. Which Means What, Exactly? *Inside Philanthropy*. https://www.insidephilanthropy. com/home/2017/4/10/mega-gifts-universities-fundraising [accessed October 8, 2018].

Siegel, D. S., and Wessner, C. 2012. Universities and the success of entrepreneurial ventures: Evidence from the small business innovation research program. *Journal of Technology Transfer*, 37(4), 404–415.

Siegel, D. S., and Wright, M. 2015. Academic Entrepreneurship: Time for a Rethink? *British Journal of Management*, 26, 582–595.

Sutter, C., Bruton, G., and Chen, J. 2018. Entrepreneurship as a solution to extreme poverty: A review and future research directions. *Journal of Business Venturing*, doi: 10.1016/j.jbusvent.2018.06.003.

The Alumni Factor. 2018. Alumni Giving. https://www.alumnifactor.com/ node/5854 [accessed October 8, 2018].

Webb, J., Ireland, D., Tihanyi, L., and Sirmon, D. 2009. You say illegal, I say legitimate: Entrepreneurship in the informal economy. *Academy of Management Review*, 34(3), 492–510.

Wennberg, K., Wiklund, J., and Wright, M. 2011. The effectiveness of university knowledge spillovers: Performance differences between university spinoffs and corporate spinoffs, *Research Policy*, 40, 1128–1143.

Wright, M. 2014. Academic Entrepreneurship, Technology Transfer and Society: Where Next?, *Journal of Technology Transfer*, 39(3), 322–334.

Wright, M. 2018. Academic Entrepreneurship: The Permanent Evolution? *Management and Organizational History*, forthcoming.

Wright, M., Clarysse, B., Lockett, A., and Knockaert, M. 2008a. Mid-range universities' in europe linkages with industry: Knowledge types and the role of intermediaries. *Research Policy*, 37, 1205–1223.

Wright, M., Lockett, A., Clarysse, B., and Binks, M. 2006. University Spin-offs and Venture Capital. *Research Policy*, 35(4), 481–501.

Wright, M., Liu, X., Buck, T., and Filatotchev, I. 2008b. Returnee entrepreneur characteristics, science park location choice and performance: An analysis of high technology SMEs in China. *Entrepreneurship Theory and Practice*, 32(1), 131–156.

Wright, M., Piva, E., Mosey, S., and Lockett, A. 2009. Business schools and academic entrepreneurship. *Journal of Technology Transfer*, 34, 560–587.

Zahra, S. and Wright, M. 2011. Entrepreneurship's next act. *Academy of Management Perspectives*, 25, 67–83.

Index

Printed in the United States
By Bookmasters